Omics in Plant Breeding

Omics in Plant Breeding

Edited by

Aluízio Borém

University of Viçosa, Viçosa, MG, Brazil

Roberto Fritsche-Neto

University of São Paulo/ESALQ, Piracicaba, SP, Brazil

WILEY Blackwell

Editorial offices: 1606 Golden Aspen Drive, Suites 103 and 104, Ames, Iowa 50010, USA
The Atrium, Southern Gate, Chichester, West Sussex, PO19 8SQ, UK
9600 Garsington Road, Oxford, OX4 2DQ, UK

For details of our global editorial offices, for customer services and for information about how to apply for permission to reuse the copyright material in this book please see our website at www.wiley.com/wiley-blackwell.

Library of Congress Cataloging-in-Publication Data has been applied for

  ISBN 978-1-118-82099-5 (paperback)

A catalogue record for this book is available from the British Library.

Cover images: iStock © pawel.gaul, iStock © Vladimirovic, iStock © emyerson

Typeset in 9.5/12.5pt Palatino by Laserwords Private Limited, Chennai, India
Printed and bound in Malaysia by Vivar Printing Sdn Bhd

1   2014

Contents

List of Contributors

Werner Camargos Antunes   Department of Biology, Maringá State University/UEM, Maringá, PR, Brazil

Francisco J.L. Aragão   Embrapa Genetic Resources and Biotechnology, Brasília, DF, Brazil

Aluízio Borém   Department of Crop Science, Federal University of Viçosa, Viçosa, MG, Brazil

Ilara Gabriela F. Budzinski   Department of Genetics, University of São Paulo/ESALQ, Piracicaba, SP, Brazil

Lucimara Chiari   Embrapa Beef Cattle, Campo Grande, MS, Brazil

Joshua N. Cobb   DuPont Pioneer, Johnston, IA, USA

Fernando Cotinguiba   Department of Genetics, University of São Paulo/ESALQ, Piracicaba, SP, Brazil

Valdir Diola (*in memoriam*)   Department of Genetics, Rural Federal University of Rio de Janeiro/UFRRJ, Seropédica, RJ, Brazil

Roberto Fritsche-Neto   Department of Genetics, University of São Paulo/ESALQ, Piracicaba, SP, Brazil

Simone Guidetti-Gonzalez   Department of Genetics, University of São Paulo/ESALQ, Piracicaba, SP, Brazil

Abdulrazak B. Ibrahim   Embrapa Genetic Resources and Biotechnology, Brasília, DF, Brazil; Department of Biochemistry, Ahmadu Bello University, Zaria, Kaduna, Nigeria; and Department of Cell Biology, University of Brasilia, DF, Brazil

Frederico Almeida de Jesus   Department of Biological Sciences, University of São Paulo/ESALQ, Piracicaba, SP, Brazil

Carlos Alberto Labate   Department of Genetics, University of São Paulo/ESALQ, Piracicaba, SP, Brazil

Mônica T. Veneziano Labate   Department of Genetics, University of São Paulo/ESALQ, Piracicaba, SP, Brazil

Marcos Antonio Machado   Department of Biotechnology, Center for Citriculture Sylvio Moreira, Agronomical Institute of Campinas, Cordeirópolis, SP, Brazil

Valéria S. Mafra   Department of Biotechnology, Center for Citriculture Sylvio Moreira, Agronomical Institute of Campinas, Cordeirópolis, SP, Brazil

Luciano Carlos da Maia   Department of Crop Science/Eliseu Maciel School of Agronomy-FAEM, Federal University of Pelotas, Pelotas, RS, Brazil

Naciele Marini   Department of Crop Science/Eliseu Maciel School of Agronomy-FAEM, Federal University of Pelotas, Pelotas, RS, Brazil

Felipe G. Marques   Department of Genetics, University of São Paulo/ESALQ, Piracicaba, SP, Brazil

Danilo de Menezes Daloso   Department of Plant Biology, Federal University of Viçosa, Viçosa, MG, Brazil; and Max-Planck-Institute for Molecular Plant Physiology, Potsdam-Golm, Germany

Hugo Bruno Correa Molinari   Laboratory of Genetics and Biotechnology, Embrapa Agroenergy, Brasília, DF, Brazil

Fabrício E. Moraes   Department of Genetics, University of São Paulo/ESALQ, Piracicaba, SP, Brazil

Ivan Miletovic Mozol   Department of Genetics, University of São Paulo/ESALQ, Piracicaba, SP, Brazil

Thiago J. Nakayama   Department of Crop Science, Federal University of Viçosa, Viçosa, MG, Brazil

Alexandre Lima Nepomuceno   Embrapa Soybean, Londrina, PR, Brazil

Antônio Costa de Oliveira   Department of Crop Science/Eliseu Maciel School of Agronomy-FAEM, Federal University of Pelotas, Pelotas, RS, Brazil

J. Miguel Ortega   Department of Biochemistry and Immunology, Federal University of Minas Gerais, Belo Horizonte, MG, Brazil

Lázaro Eustáquio Pereira Peres   Department of Biological Sciences, University of São Paulo/ESALQ, Piracicaba, SP, Brazil

Thaís Regiani   Department of Genetics, University of São Paulo/ESALQ, Piracicaba, SP, Brazil

Maria Juliana Calderan Rodrigues   Department of Genetics, University of São Paulo/ESALQ, Piracicaba, SP, Brazil

Carolina Munari Rodrigues   Department of Biotechnology, Center for Citriculture Sylvio Moreira, Agronomical Institute of Campinas, Cordeirópolis, SP, Brazil

Daniel da Rosa Farias   Department of Crop Science/Eliseu Maciel School of Agronomy-FAEM, Federal University of Pelotas, Pelotas, RS, Brazil

Janaina de Santana Borges   Department of Genetics, University of São Paulo/ESALQ, Piracicaba, SP, Brazil

Fabrício R. Santos   Department of General Biology, Federal University of Minas Gerais, Belo Horizonte, MG, Brazil

Danielle Izilda R. da Silva   Department of Genetics, University of São Paulo/ESALQ, Piracicaba, SP, Brazil

Maria Laine P. Tinoco   Embrapa Genetic Resources and Biotechnology, Brasília, DF, Brazil

Agustin Zsögön   Department of Biological Sciences, University of São Paulo/ESALQ, Piracicaba, SP, Brazil

Foreword

The application of the omics in plant breeding offers outstanding opportunities to contribute to the well being of mankind. These opportunities come about when new varieties of food, feed, fiber, and fuel crops are developed that increase productivity and confidence in the product. Such varieties have become available through traditional breeding and the use of biotechnology and they are being grown on both large and small farms. Ultimately any improved performance benefits society as a whole. Furthermore, there are good prospects for the future through the increasing opportunities associated with plant breeding, especially from the new science of omics. Many traits in the major crops, such as resistance to disease and insects, deserve more attention, and in small acreage crops plant breeding programs merit greater consideration.

This book was written to provide a broad, integrated treatment of the subjects of, for example, genomics, proteomics, metabolomics, and it relies heavily on information gleaned by the authors throughout their research careers. The fundamental principles of genetics and the background information needed for plant breeding programs are emphasized.

The intention is that the book will be used by new and advanced students, as well as serving as a reference book for those interested in the independent study of omics. Instructors are encouraged to select specific chapters to meet classroom needs depending on the desired level of teaching and the time available. Readers will also benefit from the list of references that accompany each chapter.

Aluízio Borém
Viçosa, MG, Brazil
and
Roberto Fritsche-Neto
Piracicaba, SP, Brazil
Editors

1 Omics: Opening up the "Black Box" of the Phenotype

Roberto Fritsche-Neto[a] and Aluízio Borém[b]

[a]Department of Genetics, University of São Paulo/ESALQ, Piracicaba, SP, Brazil

[b]Department of Crop Science, Federal University of Viçosa, Viçosa, MG, Brazil

From the time that is believed agriculture began, in approximately 10 000 BC, people have consciously or instinctively selected plants with improved characteristics for cultivation of subsequent generations. However, there is disagreement as to when plant breeding became a science. Plant breeding became a science only after the rediscovery of Mendel's laws in 1900. However some scientists disagree with this view. It was only in the late 19th century that the monk Gregor Mendel, working in Brno, Czech Republic, uncovered the secrets of heredity, thus giving rise to genetics, the fundamental science of plant breeding.

Scientists added a few more pieces to the puzzle that was becoming this new science in the first half of the 20th century by concluding that something inside the cells was responsible for heredity. This hypothesis generated answers and thus consequent new hypotheses, leading to the continuing accumulation of knowledge and progress in the field. For example, the double helix structure of DNA was elucidated in 1953 (Table 1.1). Twenty years later, in 1973, the first experience with genetic engineering opened the doors of molecular biology to scientists. The first transgenic plant, in which a bacterial gene was inserted stably into a plant genome, was produced in 1983. Based on these advances, futuristic predictions about the contribution of biotechnology were published in the media, both by laypeople and scientists themselves, creating great expectations for its applications. Euphoria was the tone of the scientific community. Many companies, both large and small, were created, encouraged by the prevailing enthusiasm of the time (Borém and Miranda, 2013).

Many earlier predictions have now become reality (Table 1.1), leading to the consensus that each year the benefits of biotechnology will have a greater impact on breeding programs. Consequently, new companies

Table 1.1   Chronology of major advances in genetics and biotechnology relevant to plant breeding. Adapted from Borém and Fritsche-Neto (2013).

Year	Historical landmark
1809–1882	Charles Darwin develops the theory of natural selection: those individuals most adapted to their environment are selected, survive, and produce more offspring.
1865	Gregor Mendel establishes the first statistical methodologies applicable to plant breeding, giving rise to the "era of genetics," with his studies on the traits of pea seeds.
1910	Thomas Morgan, studying the effects of genetic recombination in *D. melanogaster*, demonstrates that genetic factors (genes) are located on chromosomes.
1941	George Beadle and Edward Tatum demonstrate that a gene produces a protein.
1944	Barbara McClintock elucidates the process of genetic recombination by studying satellite chromosomes and genetic crossing-over related to linkage groups in chromosomes 8 and 9 of maize.
1953	James Watson and Francis Crick, using X-ray diffraction, propose the double helix structure of the DNA molecule.
1957	Hunter and Markert develop biochemical markers based on the expression of enzymes (isoenzymes).
1969	Herbert Boyer discovers restriction enzymes, opening new perspectives for DNA fingerprinting and the cloning of specific regions.
1972	Recombinant DNA technology begins with the first cloning of a DNA fragment.
1973	Stanley Cohen and Herbert Boyer perform the first genetic engineering experiment on a microorganism, the bacterium *Escherichia coli*. The result was considered to be the first genetically modified organism (GMO).
1975	Sanger develops DNA sequencing by the enzymatic method; in 1984, the method was improved, and the first automatic sequencers were built in the 1980s.
1977	Maxam and Gilbert develop DNA sequencing by chemical degradation.
1980	Botstein *et al.* develop the RFLP (Restriction Fragment Length Polymorphism) technique for genotypic selection.
1983	The first transgenic plant is produced, a variety of tobacco into which a group of Belgian scientists introduced kanamycin antibiotic resistance genes.
1985	Genentech becomes the first biotech company to launch its own biopharmaceutical product, human insulin produced in cultures of *E. coli* transformed with a functional human gene.
1985	The first plant with a resistance gene against Lepidoptera is produced.
1986	The first field trial of transgenic plants is conducted in Ghent, Belgium.
1987	The first plant tolerant to a herbicide, glyphosate, is created.
1987	Mullis and Faloona identify thermostable Taq DNA polymerase enzyme, which enabled the automation of PCR.
1988	The first transgenic cereal crop, Bt maize, is developed.
1990	Rafalski *et al.* (1990) develop the first genotyping technique using PCR, RAPD (Random Amplified Polymorphism DNA).

Table 1.1   *(continued)*

Year	Historical landmark
1990	New tools for NCBI sequence alignment are created (BLAST – Basic Local Alignment Search Tool) (National Center of Biotechnology and Information – www.ncbi.gov).
1994	The first permit is issued for the commercial cultivation and consumption of a GMO, the Flavr Savr tomato.
1997	The first plant containing a human gene, the human protein c-producing tobacco, is produced.
2000	The first complete sequencing of a prokaryotic organism, the bacterium *E. coli*, is conducted.
2003	The first eukaryotic genome sequence, that of the human, is released by two major independent research groups in the United States.
2005	Large-scale sequencing (NGS – Next Generation Sequencing) is used as a tool to unravel whole genomes quickly.
2011	Second- and third-generation large-scale sequencing systems are developed; eukaryotic genomes are sequenced in just a few hours.
2012	Technologies that control the temporal and spatial expression of genes are used in genetic transformation and the exclusion of auxiliary genes.
2014 ...	Large-scale sequencing, macro- and microsynteny, associative mapping, molecular markers for genomic selection, QTL (quantitative trait loci) cloning, and large-scale phenotyping are widely used, the use of "omics" and specific, multiple GMO phenotypic traits is expanded and bioinformatics is used intensively.

have been established to take advantage of innovative, highly promising business opportunities.

The Post-Genomics Era

In the late 20th and early 21st centuries, genome sequencing studies developed rapidly. Gene sequences are now available for entire organisms, including humans. After these DNA base sequences are determined, it is necessary to organize them and identify the coding regions and their functions in the organism.

In this context, with a huge range of sequences being deposited in databases, geneticists are faced with a challenge as great as that which propelled the "genomics era": correlating structure with function. This challenge has given rise to functional genomics, the science of the "era of omics."

Omics is the neologism used to refer to the fields of biotechnology with the suffix omics: genomics, proteomics, transcriptomics, metabolomics, and physiognomics, among others. These new tools are helping to

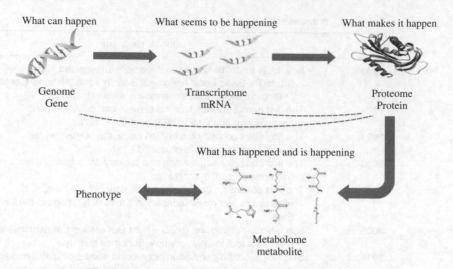

Figure 1.1   Systems biology: from genome to phenotype.

develop superior cultivars for food production or even allowing plants to function as biofactories. The focal point for the 21st century will be the technological development of large-scale molecular studies and their integration into systems biology. These studies aim to understand the relationship between the genome of an organism and its phenotype, that is, to open up the "black box" that contains the path between codons and yield or resistance to biotic or abiotic stresses (Figure 1.1). Thus, systems biology is a science whose objectives are to discover, understand, model, and design the dynamic relationships between the biological molecules that make up living beings to unravel the mechanisms controlling these parts.

The Omics in Plant Breeding

In recent years, genetics and omics tools have revolutionized plant breeding, greatly increasing the available knowledge of the genetic factors responsible for complex traits and developing a large amount of resources (molecular markers and high-density maps) that can be used in the selection of superior genotypes. Among the existing omics tools, global transcriptome, proteome, and metabolome profiles created using EST, SAGE, microarray, and, more recently, RNA-seq libraries have been the most commonly used techniques to investigate the molecular basis of the responses of plants, tissues/organs or developmental stages to experimental conditions (Kumpatla *et al.*, 2012). However, regardless of the omics used, the aid of bioinformatics is required for the analysis and interpretation of the data obtained.

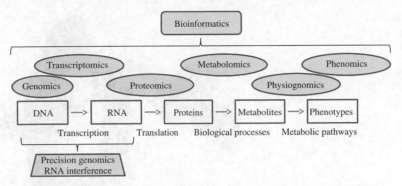

Figure 1.2   Trail of omics and the relationships among the fields and their corresponding biological processes.

Given the importance of these fields, subsequent chapters will discuss the various tools currently in use, or with great potential for future use, in plant breeding. The roles of these fields, the relationships between them and their corresponding biological processes (as well as their presentation in this book) can be visualized by the "trail" of omics, as shown in Figure 1.2.

The initial draft of the genome of the first plant to be sequenced (*Arabidopsis thaliana*) took approximately ten years to be developed. Today, with the use of the next generation of DNA sequencing technology (NGS, Next-Generation Sequencing) (e.g., Oxford Nanopore, PacBio RS, Ion Torrent, and Ion Proton, among others) and powerful bioinformatics and computational modeling programs, genomes can be sequenced, assembled, and related to the phenotypic traits specific to each genotype within a few weeks. This capability, combined with the drastic reduction in the cost of sequencing, has enabled the generation of an ever-increasing volume of data, thus enabling the comprehensive study of genomes and the development of informative molecular markers.

Genomics, Precision Genomics, and RNA Interference

All of this information has inspired the development of new strategies for genetic engineering. However, until recently, the available genetic engineering tools could only introduce changes into larger blocks of DNA sequences, which could subsequently be inserted only at random in the genome of a target species. Recent advances in this field have made it possible to obtain new variations from site-directed modifications, including specific mutations, insertions, and substitutions of genes and/or blocks of genes, making genetic engineering a precise and powerful alternative for the development of new cultivars.

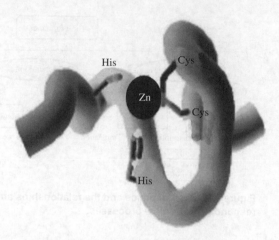

Figure 1.3   Zinc finger nucleases (ZFNs).

These modifications to specific DNA sequences are initiated by generating a break on the double-stranded target DNA (Double Stranded DNA Break, DSB). Genetically modified nucleases are designed to identify the specific site of the target genome and catalyze the creation of the DSB, enabling the desired DNA modifications to occur at the specific break site or close to it.

To access specific sites, three enzymes have been genetically modified or constructed: zinc finger nucleases (ZFNs) (Figure 1.3), transcription activator-like effector nucleases (TALENs) (Figure 1.4) and meganucleases, also known as LAGLIDADG hormone endonucleases (LHEs).

Another widely used technique is post-transcriptional gene silencing (PTGS), or RNA interference (RNAi). This technique has assisted the development of transgenic plants capable of suppressing the expression of endogenous genes and foreign nucleic acids (Aragão and Figueiredo, 2008).

Knowledge about the mechanisms involved in RNA-mediated gene silencing has been important in the understanding of the biological function of genes, the interaction between organisms, and the development of new cultivars, among other applications.

The RNAi pathway begins with the presence of double-stranded RNA (dsRNA) in the cytoplasm, which may vary in origin and size (Figure 1.5). These dsRNAs are cleaved by the Dicer enzyme, a member of the RNase III nuclease family. After the processing of the dsRNA, small interfering RNAs (siRNAs) are formed, which are then integrated into an RNA-induced silencing complex (RISC). The RISC is responsible for the cleavage of a specific mRNA target sequence.

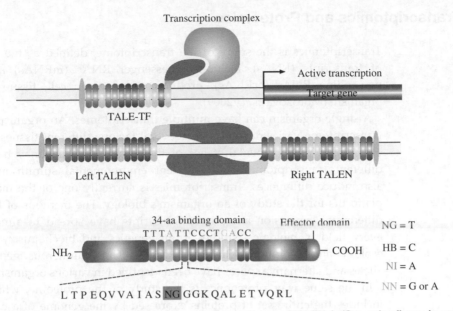

Figure 1.4   Transcription activator-like effector nucleases (TALENs). (See color figure in color plate section).

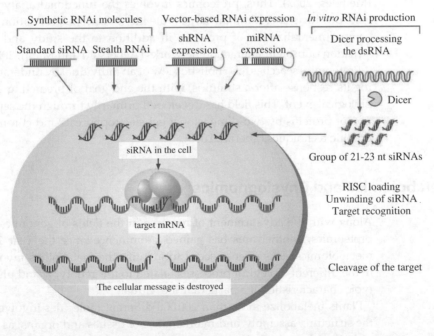

Figure 1.5   Pathways of gene silencing in plant cells. (Source: Based on Souza *et al.*, 2007). (See color figure in color plate section).

Transcriptomics and Proteomics

Transcriptomics is the study of the transcriptome, defined as the set of transcripts (RNAs), including messenger RNAs (mRNAs) and non-coding RNAs (ncRNAs), produced by a given cell, tissue or organism (Morozova *et al.*, 2009).

A single organism can have multiple transcriptomes. An organism's transcriptome varies depending on several factors: different tissues or organs and developmental stages of the same individual may have different transcriptomes, and different environmental stimuli may also induce differences. Transcriptomics is currently one of the main platforms for the study of an organism's biology. The methods of the differential expression analysis of transcripts have spread to almost every field of biological studies, from genetics and biochemistry to ecology and evolution (Kliebestein, 2012). Thus, numerous genes, alleles and alternative splices have been identified in various organisms.

In the same way, proteomics is the study of the proteome, which includes the entire set of proteins expressed by the genome of a cell, tissue or organism. However, this study can be directed only to those proteins that are expressed differentially under specific conditions (Meireles, 2007). Thus, proteomics involves the functional analysis of gene products, including the large-scale identification, localization, and compartmentalization of proteins, in addition to the study and construction of protein interaction networks (Aebersold and Mann, 2003).

Proteomics searches for a holistic view of an individual by understanding its response after a stimulus, with the end goal of predicting some biological event. This field has developed primarily through the separation of proteins by two-dimensional gel electrophoresis and chromatographic techniques (Eberlin, 2005).

Metabolomics and Physiognomics

Along with the advancement of research in the fields of genomics and proteomics, another area has gained prominence since the year 2000: metabolomics. This science seeks to identify the metabolites involved in the different biological processes related to the genotypic and phenotypic characteristics of a particular individual.

Plants metabolize more than 200 000 different molecules involved in the structure, assembly, and maintenance of tissues and organs, as well as in the physiological processes related to growth, development, and reproduction. Metabolic pathways are complex and interconnected, and they are, to some degree, dependent on and regulated by their own products or substrates, as well as by their genetic components and different levels of gene regulation.

This observation shows the great capacity for modulation or plasticity of the physiological response networks of plants under the same hierarchical control (DNA). Through the combined and simultaneous analysis of more than one regulatory level, such as the association of molecular markers and metabolic comparisons, a complex set of data can be generated, that is, the physiognomy. This science, in turn, generates systemic models aiming to understand and predict plant responses to certain stimuli and/or environmental conditions.

Phenomics

The field of phenomics employs a series of "high-throughput" techniques to enhance and automate the ability of scientists to accurately evaluate phenotypes, as well as to eventually reduce the determinants of phenotype to genes, transcripts, proteins, and metabolites (Tisné *et al.*, 2013).

The phenome of an organism is dynamic and uncertain, representing a set of complex responses to endogenous and exogenous multidimensional signals that have been integrated during both the evolutionary process and the developmental history of the individual. This phenotypic information can be understood as a set of continuous data that change during the development of the species, the population, and the individual in response to different environmental conditions.

The emphasis of phenomics is phenotyping in an accurate (able to effectively measure characteristics and/or performance), precise (little variance associated with repeated measurements), and relevant manner within acceptable costs. This focus is important because phenotyping is currently the main limiting factor in genetic analysis. Unlike genotyping, which is highly automated and essentially uniform across different organisms, phenotyping is still a manual, organism-specific activity that is labor intensive and is also very sensitive to environmental variation.

The following are examples of phenomics approaches: (i) the use of digital cameras to take zenithal images for the automatic analysis of leaf area and rosette growth and the measurement of the characteristics of tissues, organs or individuals (Tisné *et al.*, 2013); (ii) the use of infrared cameras to visualize temperature gradients, which can indicate the degree of energy dissipation (Munns *et al.*, 2010) and have implications for responsiveness to drought stress and photosynthetic rate; (iii) the use of images generated by fluorescence detectors to identify the differential responses of populations of seedlings, fruits or seeds to a stressor (Jansen *et al.*, 2009); (iv) the use of noninvasive methods to visualize subterranean systems (Nagel *et al.*, 2012); and (v) the use of LIDAR (Light Detection and Ranging) technology to measure growth rate through differences between small distances measured using a laser (Hosoi and Omasa, 2009).

All these instruments generate objective digital data that can be transmitted to remote servers, many of which are connected to the Internet, for storage and further analysis, which is also often automated. The prospects for these technologies are very promising for breeding programs, which increasingly evaluate greater numbers of individuals.

Bioinformatics

The exponential increase in the volume of both molecular and phenotypic data requires increased computational capacity for its storage, processing, and analysis. To this end, numerous computers and analytical tools have been developed to address the massive volume of data originating from genomics, proteomics, metagenomics, and metabolomics, among other omics.

Biological data are relatively complex compared with those from other scientific fields, given their diversity and their interrelationships. All this information can only be organized, analyzed, and interpreted with the support of bioinformatics.

Bioinformatics can be defined as the field that covers all aspects of the acquisition, processing, storage, distribution, analysis, and interpretation of biological information. A number of tools that aid in understanding the biological significance of omics data have been developed through the combination of procedures and techniques from mathematics, statistics, and computer science. In addition, the creation of databases with previously processed information will accelerate research in other biological fields, such as medicine, biotechnology, and agronomy.

Prospects

Plant breeding is an art, science, and business that is little more than a century old. Using methods developed mainly in the 20th century, breeders have developed agronomically superior cultivars. Because of the constantly increasing challenge of agricultural food production, plant breeding must evolve and use new knowledge. Therefore, the omics will gradually assume greater relevance and be incorporated into the routines of breeding programs, making them more accurate, fast, and efficient. Although the challenges are great, the prospects are even greater.

References

Aebersold, R.H.; Mann, M. 2003. Mass spectrometry-based proteomics. Nature, 422 (6928): 198–207.

Aragão, F.J.L; Figueiredo, S.A. 2008. RNA interference as a tool for plant biochemical and physiological studies. In: Rivera-Domínguez, M., Rosalba-Troncoso, R., and Tiznado-Hernández, M.E. (eds.) A Transgenic Approach in Plant Biochemistry and Physiology. Kerala, India: Research Signpost, pp. 17–50.

Borém, A.; Fritsche-Neto, R. 2013. Biotecnologia aplicada ao melhoramento de plantas. 6th edn. Visconde do Rio Branco: Suprema Publishers, 336 pp.

Borém, A.; Miranda, G.V. 2013. Melhoramento de plantas. 6th edn. Viçosa: UFV Publishers, 523 pp.

Eberlin, M. 2005. A proteômica e os novos paradigmas. Reportagem Jornal da Unicamp, Universidade Estadual de Campinas. UNICAMP. p. 3. 14–27 November, 2005.

Hosoi, F.; Omasa, K. 2009. Detecting seasonal change of broad-leaved woody canopy leaf area density profile using 3D portable LIDAR imaging. Functional Plant Biology, 36 (11): 998–1005.

Jansen, M.; Gilmer, F.; Biskup, B.; *et al.* 2009. Simultaneous phenotyping of leaf growth and chlorophyll fluorescence via GROWSCREEN FLUORO allows detection of stress tolerance in *Arabidopsis thaliana* and other rosette plants. Functional Plant Biology, 36 (11): 902–914.

Kliebenstein, D.J. 2012. Exploring the shallow end; estimating information content in transcriptomics studies. Frontiers in Plant Science, 3: 213.

Kumpatla, S.P.; Buyyarapu, R.; Abdurakhmonov, I.Y.; Mammadov, J.A. 2012. Genomics-assisted plant breeding in the 21st century: Technological advances and progress. In: Abdurakhmonov, I. (ed.). Plant Breeding. Rijeka, Croatia: InTech. pp. 131–184.

Meireles, K.G.X. Aplicações da Proteômica na Pesquisa Vegetal. Campo Grande, MS: Embrapa. Document 165. ISSN 1983-974X. September, 2007.

Morozova, O.; Hirst, M.; Marra, M.A. 2009. Applications of new sequencing technologies for transcriptome analysis. Annual review of genomics and human genetics, 10:135–151.

Munns, R.; James, R.A.; Sirault, X.R. *et al.* 2010. New phenotyping methods for screening wheat and barley for beneficial responses to water deficit. Journal of Experimental Botany, 61 (13): 3499–3507.

Nagel, K.A.; Putz, A.; Gilmer, F.; *et al.* 2012. GROWSCREEN-Rhizo is a novel phenotyping robot enabling simultaneous measurements of root and shoot growth for plants grown in soil-filled rhizotrons. Functional Plant Biology, 39 (11): 891–904.

Rafalski, J.A., Tingey, S.V., Williams, J.G.K., 1991. RAPD markers, a new technology for genetic mapping and plant breeding. AgBiotech News and Information, 3: 645–648.

Souza, A.J.; Mendes, B.M. J.; Filho, F.A.A.M. 2007. Gene silencing: concepts, applications, and perspectives in wood plants. Scientia Agricola (Piracicaba, Braz.), 64: 645–656.

Tisné, S.; Serrand, Y.; Bach, L.; *et al.* 2013. Phenoscope: an automated large-scale phenotyping platform offering high spatial homogeneity. The Plant Journal, 74 (3): 534–544.

2 Genomics

Antônio Costa de Oliveira, Luciano Carlos da Maia,
Daniel da Rosa Farias, and Naciele Marini
*Department of Crop Science/Eliseu Maciel School of
Agronomy-FAEM, Federal University of Pelotas, Pelotas, RS, Brazil*

The Rise of Genomics

The constant technological advances of modern societies, evident in, for example, the use of synthetic oil based commodities, still requires many products from plants in general domestic use. Since early civilizations, men have tried to adapt plants to their needs. Improved techniques have evolved significantly since the 19th century, when directed crossings were developed. These were followed by the rediscovery of the genetic principles that had been established by Mendel early in the 20th century, reaching the green revolution in the 1960s (Borlaug, 1983; Allard, 1971) and finally the gene revolution of 1990–2000s, as discussed in Bologna (Tuberosa, personal communication). Thus, throughout the whole of the 20th century, advances in genetics allowed systematic initiatives towards the increase in food production and vegetable fibers, in addition to other edible/inedible byproducts. However, there is still a great need to improve yields in order to cope with rises in food demand without increasing the areas of land that are cultivated. This balance between food supply and food demand has been maintained by breeders to date, but population increases still present a challenge (Food and Agricultural Organization of the United Nations (FAO), 2010). For practical purposes, the focus of this chapter will be restricted to the description and use of genomics in plant breeding.

Plant genomics consists of the development of large-scale analyses of structural and functional features of genomes, allowing the discovery of evolutionary and functional dynamics in plants.

DNA Sequencing

DNA sequencing was first performed in the 1970s with two different techniques, the Sanger chemistry using dideoxy chain termination

Omics in Plant Breeding, First Edition. Edited by Aluízio Borém and Roberto Fritsche-Neto.
© 2014 John Wiley & Sons, Inc. Published 2014 by John Wiley & Sons, Inc.

(Sanger, Nicklen, and Coulson, 1977), and also chemical degradation (Maxam and Gilbert, 1977). These pioneering efforts brought about spectacular results but at a slow pace, since the sequencing was performed manually and less than 200 bases were produced during the reading of four lanes. An automatic sequencer was released in the 1980s, using the Sanger chemistry (Connell *et al.*, 1987), allowing more significant advances with a high throughput. DNA sequencing represents a powerful tool for the identification of a wide range of biological phenomena through the collection of large sets of data samples. This strategy starts with DNA extraction, which can be at the whole genome level (Shotgun sequencing) or based on smaller fractions cloned in BACs (Bacterial Artificial Chromosomes or YACs (Yeast Artificial Chromosomes). These large vectors are subsequently fragmented into smaller plasmid/cosmid libraries by means of enzyme digestion or other fractionation methods. The ends of smaller clones are then sequenced and read in the sequencing platforms. The data are generated in FASTA format and bioinformatic algorithms are used to build contigs and scaffolds, therefore reconstructing the original DNA molecule (chromosome). This task is hierarchical, as it goes from smaller to larger sequences, and is performed with the aid of assembly software, generating contigs (generated from overlapping sequence reads) and scaffolds (generated from the ordering of overlapping contigs) (Figure 2.1).

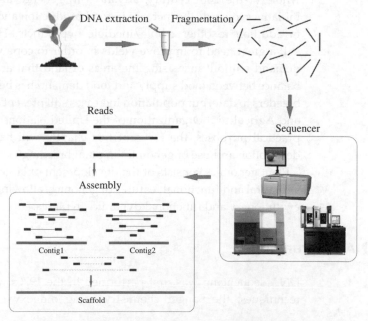

Figure 2.1   Basic scheme, representing the process of plant DNA sequencing. (See color figure in color plate section).

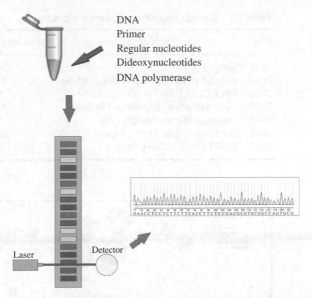

DNA
Primer
Regular nucleotides
Dideoxynucleotides
DNA polymerase

Laser   Detector

GAACCTCCTCTTCTTCACCTTCTCCGACGCGTGCGGCCAGTGCG

Figure 2.2   Sanger automatic sequencing. The process initiates with a PCR in which, besides DNA, primers and deoxynucleotides, dideoxyribonucleotides (ddNTP) are added, which are labeled with fluorescent markers and without hydroxils at the 3'-position. Thus, every time a ddNTP is added to the chain, its extension is interrupted and, after numerous cycles, many different size fragments will be generated, which have been terminated by different ddNTPs. These fragments are later subjected to electrophoresis and separated into their different sizes. The fragments pass through a laser and the fluorescence is detected; this signal is then transformed into a chromatogram (a graph with peaks of different colors), each colored peak being attributed to a different base. The sequencer comes with software that will convert chromatograms into FASTA format sequences. (See color figure in color plate section).

The first genome sequencing projects were performed using automatic Sanger sequencing chemistry (Figure 2.2). In the period between 1995 and 2005, the genomes of *Arabidopsis* (*Arabidopsis* Genome Initiative, 2000) and rice (International Rice Genome Sequencing Project, 2005) were completed.

Since 2005, a new generation of technologies has emerged (Varshney *et al.*, 2009), allowing the generation of an even larger amount of data per sample by reaction and revolutionizing the genomics landscape (Table 2.1). These technologies include the advances made in nanobiology and robotics and contributed to one of the goals in the world of science, that is, to perform the sequence of any human genome for a price of under US $1000 dollars (Thudi *et al.*, 2012).

All these technologies perform DNA sequencing on platforms capable of generating several million base outputs in a single run. Among the NGS (Figure 2.3), two are in use worldwide: the Roche 454 FLX and Solexa/Illumina.

Table 2.1 Sanger and NGS sequencing platforms.

Year	Technology	Read length (bases)	Bases per run
1977	Sanger	1000	100 Kb
2005	454 (Life Science/Roche Diagnostics)	500	500 Mb
2005	ABI SOLID (Life Technologies)	50	30 Gb
2007	Illumina Genome Analyser (Solexa)	150	300 Gb
2010	Helicos (Helicos Biosciences)	55	35 Gb
2010	Ion Torrent (Life Technologies)	200	1 Gb
2010	SMRT (Pacific Biosystems)	2000	100 Mb

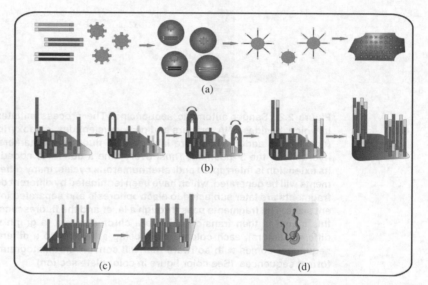

Figure 2.3 NGS clone amplification: (a) emulsion PCR, used by 454, Polonator, and ABI SOLID platforms; (b) bridge amplification used by Illumina; (c) single molecule sequencing used by Helicos; and (d) real time sequencing, used by SMRT. (Source: Adapted from Metzker, 2009; Shendure and Ji, 2008). (See color figure in color plate section).

The strategy used by 454, Polonator, and ABI SOLID consists of an emulsion PCR (Dressman *et al.*, 2003). The DNA fragments are linked to adaptors, which are linked to magnetic microbeads by means of pairing between the adaptors and complementary oligonucleotide sequences present in the bead surface. Only one type of fragment pairs with one bead. The microbeads are individually captured in oil drops, where the emulsion PCR occurs. Thousands of copies from the target fragment are produced and the microbeads containing the target sequences are subsequently captured in wells where the pyrosequencing occurs. Each base is identified by means of specific chemiluminescence emission. In the Illumina platform, adaptors are linked to both ends of a DNA fragment. DNA molecules are then attached to a solid matrix containing oligonucleotides complementary to the adaptors. Thus, the fragment free adaptor end is linked to the complementary oligonucleotide in the matrix and a bridge structure is formed. Next, a PCR reaction is performed, with the addition of labeled nucleotides. After a series of annealing cycles, bridge formation, amplification, and denaturation, clusters of identical molecules are formed attached to the matrix, and the nucleotide reading is performed sequentially through the signal emitted by a laser beam. The single molecule sequencing of the Helicos platform is initiated with the addition of a polyA tail into DNA fragments. These fragments are then hybridized to a solid matrix that contains thousands of adhered polyT fragments, and in each cycle a unique fluorescence labeled fragment complementary to the template DNA is added, a camera then captures this fluorescence and an image is produced for the nucleotide reading. Real time sequencing, performed by SMRT, is obtained by individual molecules of DNA polymerase adherent to the bottom of a surface, known as a Zero Mode Waveguide (ZWM), which performs the reading of each base added to the template strand of DNA. These nucleotides are marked with a fluorescent label and the readings are taken by detecting each incorporated nucleotide fluorescence signal (Metzker, 2009; Shendure and Ji, 2008).

Owing to reductions in cost and an increase in sequencing capacity, the new platforms are efficient for routine use in sequencing and resequencing of individual genomes, for detecting variations between target and reference genomes (Service, 2006). In the last few years, NGS technologies have rapidly evolved, with the potential to accelerate the discoveries in biological and medical research, stimulating gene variation studies, understanding of biotic and abiotic stress responses and evolutionary studies (Shendure and Ji, 2008). Resequencing whole genomes is an essential tool for the characterization of genetic variations in many contexts (Bentley, 2006). These sequencing platforms are being used in large plant genome sequencing projects such as IOMAP – International *Oryza* Map Genome Initiative, which aims to sequence all species from the *Oryza* genus, the "1000 Plant Genomes Project" (www.onekp.com/), the "1001 *Arabidopsis* Genome Project" (www.1001genomes.org/), and the "1000 Plant and Animal Genome Project" (www.1d1.genomics.cn/). Additionally, the "Genome 10K Project" was created to sequence and assemble 10 000 vertebrate genomes, including at least one from each genus (www.genome10k.org/).

As a result, the new sequencing platforms generate short reads, which are smaller than those generated by traditional Sanger sequencing, and the assembly of these reads into contigs has become one of the largest challenges for these technologies (Figure 2.4). Therefore, bioinformatics has developed into a key tool for biological, genetic, and plant breeding studies.

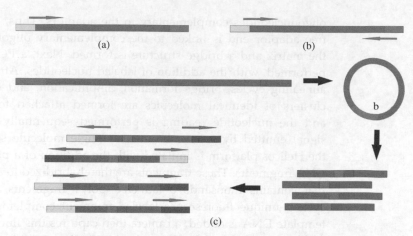

Figure 2.4 Reads generated by NGS. (a) Single-end reads, where the reading is performed only on one end (indicated by the blue arrow). (b) Paired-end reads, where the sequencing is performed in both fragment ends, in opposite directions (blue arrows). (c) In the mate-pair reads, biotin-labeled nucleotides are linked to both ends of the fragment, which is circularized and cut in small sequences that are selected by biotin-based rescue and moved to sequence reading (single by 454 and paired by Illumina and SOLID. (Source: Adapted from Hamilton and Buell, 2012). (See color figure in color plate section).

During the last two decades, the major advances in genome technology have lead to an increase in the amount of biological information generated by the scientific community. Thus, storage of biological data in public databases is becoming more common every year, with an exponential growth in size (Figure 2.5).

Development of Sequence-based Markers

The treatment of genetic information began to increase in scale from the 1980s when the first markers were generated (Botstein *et al.*, 1980). With the development of automated sequencing technologies, new markers were made available with larger outputs. Therefore, bioinformatic tools became part of genomic research and a routine procedure in genomics and breeding laboratories.

Currently, the combination of novel genomics insights with traditional breeding methods is seen as essential to achieving progress in agriculture. Genome access became a fundamental resource in biological sciences and, although the rate of sequencing in plants is slower compared with microbial and mammal systems, genomics has been largely applied in agronomy, biochemistry, forest sciences, genetics, horticulture, plant pathology, and systematics (Hamilton and Buell, 2012). Therefore, the term molecular breeding was coined to describe the integration of classic

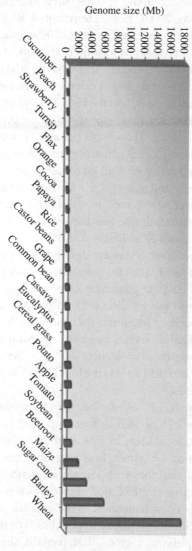

Figure 2.5 Genome size of cultivated species. Published data (gray) and four economically important species (orange). (See color figure in color plate section).

breeding and molecular biology, for example, molecular markers (Xu, Li, and Thomson, 2012).

Besides the contribution given by molecular markers to plant breeding advances, the increase in availability of data referring to sequenced genomic regions and transcriptome analyses have also played an important role (Varshney *et al.*, 2009). Complete or partial genomes are available for many species, such as *Arabidopsis* (Arabidopsis Genome Initiative, 2000), rice (International Rice Genome Sequencing Project, 2005), ice-plant (Tuskan, *et al.*, 2006), grape (Jaillon, *et al.*, 2007), papaya (Ming *et al.*, 2008), sorghum (Paterson, *et al.*, 2009), *Brachypodium* (Vogel *et al.*, 2010) and cocoa (Argout *et al.*, 2011).

The understanding of reference genomes, which are available for several plant species, and the high throughput nature of sequencing projects have provided new opportunities for plant breeding, such as comparative genomics. These tools became very useful thanks to the new experimental and bioinformatics approaches for data treatment. The future of crop improvement will be centered on comparisons of individual plant genomes, and some of the best opportunities may lie in using combinations of new genetic mapping strategies and evolutionary analyses to direct and optimize the discovery and use of genetic variation (Morrel, Buckler, and Ross-Ibarra, 2012).

Molecular markers are defined as genetic markers that are based on a pool of techniques that can detect DNA polymorphism in a single locus or in the whole genome (Oliveira, 1998; Varshney, Graner, and Sorrells, 2005) and can be detected by Polymerase Chain Reaction (PCR) methods or other procedures that combine the use of restriction enzymes and hybridization techniques between complementary DNA sequences.

Among these molecular markers, Restriction Fragment Length Polymorphism (RFLP) was the first DNA marker to be used in plant breeding (Figure 2.6).

After RFLP, a number of PCR-based markers were generated (Figure 2.7), as well as the Amplified Fragment Length Polymorphism (AFLP) technique. In this technique, the DNA is fragmented with two restriction enzymes, one rare and one frequent cutter, followed by linking adaptors with known sequences linked to both fragment ends. A selective amplification of fragments is next performed with the use of primers complementary to the adaptor's sequences.

The Simple Sequence Repeat (SSR) or microsatellite technique consists in amplifying regions that contain simple tandem repeats, with the aid of primers complementary to the unique regions that flank the repeat region.

Single Nucleotide Polymorphism (SNP) markers have great potential in genomics and breeding studies, especially with the increase in availability of sequencing data. They can either be detected by a DNA labeled probe that amplifies a specific allele or by sequencing fragments from

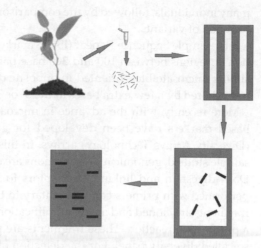

Figure 2.6   Scheme representing RFLP. The DNA is extracted and fragmented and the fragments are separated by electrophoresis in agarose gels, where a continuous smear is seen. The fragments are denatured and transferred to nylon or nitrocellulose membranes, followed by hybridization with a radioactive fluorescent probe containing a single-strand target fragment. Autoradiography, in which the membrane hybridized with the labeled probe is exposed to an X-ray film. (See color figure in color plate section).

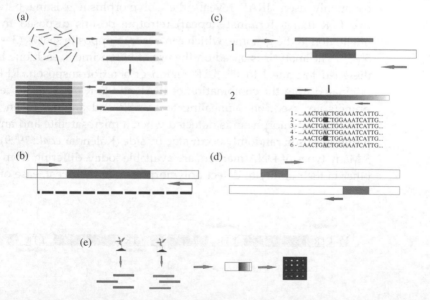

Figure 2.7   Molecular marker depictions: (a) AFLP; (b) SSR; (c) SNP; (d) ISSR; and (e) DArT. (See color figure in color plate section).

many individuals, followed by the comparison of sequences for the identification of variants.

Inter-Simple Sequence Repeat (ISSR) markers consist in the amplification of regions between 100 and 300 base pairs between two identical/similar microsatellites, oriented in opposite directions. The primers can be anchored by a few extra bases at the 5' or 3' end of the repeat.

More recently, with the advance in microarray research, microarray-based markers have been developed for genotyping, such as DArT (Diversity Arrays Technology) arrays. In this technique, for each DNA sample studied, genomic representations are developed, generated from DNA digestion and linking of adaptors to fragments. Then a PCR is performed with primers complementary to the adaptors, the amplified fragments are cloned and a new amplification is performed followed by a purification reaction. These fragments are placed on a solid matrix, a so-called diversity panel. For the detection of polymorphism, representations prepared from individual target genomes are hybridized to the diversity panel. The polymorphic clones show variable hybridization signal intensities for different individuals (Jaccoud *et al.*, 2001).

Another class of markers used in polymorphism detection is composed of those based in transposon amplification. Inter-Retrotransposon Amplified Polymorphism (IRAP) and Retrotransposon-Microsatellite Amplified Polymorphism (REMAP) (Figure 2.8) are among those most commonly used. IRAP reveals the polymorphism existing between two LTR (Long Terminal Repeat) retrotransposons disposed in any orientation in the genome, which can occur in opposite senses (3'–5' or 5'–3'). The analysis is based on the use of two primers, each one being designed to anneal to the LTR region of a retrotransposon. REMAP originated from the combination of IRAP and SSR markers, since two primers are used, one annealing to an SSR and another to an LTR. Thus, the polymorphism is detected when a microsatellite and an LTR retrotransposon randomly occur side by side (Kalendar *et al.*, 1999).

Many types of DNA markers are available today, differing from each other in their ability to detect polymorphism, in their cost, ease of use,

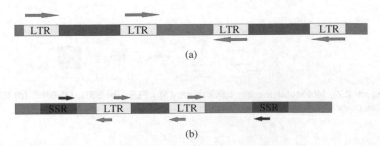

(a)

(b)

Figure 2.8   IRAP (a) and REMAP (b) marker techniques. (See color figure in color plate section).

and robustness. These techniques have been used in genetic diversity, fingerprinting, seed genetic purity analyses, genetic mapping, gene isolation, and assisted breeding (Borém, 2009). Among the diverse applications in plant breeding, one can highlight marker-assisted breeding, which is fundamental to the mapping and association of markers to genes that control important traits (Oliveira, 1998). Currently, the advances in sequencing technologies supply efficient information for the large-scale generation of markers that can be used in plant breeding programs (Henry *et al.*, 2012).

Genetic Mapping

A linkage map consists of a linear representation of a group of molecular markers in a linkage cluster. These markers can be genes with known functions, chromosomal regions or DNA fragments identified by some specific technique (Figure 2.9).

Genetic maps have been the object of studies since the early decades of the 20th century. The limitation found by pioneer geneticists to construct genetic maps was the inadequate number of morphological markers. However, for some species, the occurrence of mutations allowed the construction of the first low-density genetic maps (Oliveira, 1998).

The initial ideas on genetic mapping emerged with papers by Morgan and colleagues. They revealed the first evidence that genes were located in definite positions on chromosomes and could be experimentally manipulated and evaluated (Rocha *et al.*, 2003). It was Sturtevant, in 1913, who suggested the use of recombinant frequency as a measure of distance between genes, allowing the construction of maps in which the relative position of phenotypic markers in a given organism could be presented. These ideas slowly developed through the decades of the 1920s, 1930s, and 1940s, with the input from many scientists, such as R.A. Fisher, J.B.S. Haldane, D.D. Kosambi, A.R.G. Owen, among others (Bailey, 1961). However, the discovery of DNA markers revolutionized the field, making the construction of enriched/saturated maps feasible.

Molecular mapping is performed with molecular biology techniques and a population according to the principles of Mendelian laws. As such, it is necessary to use plants with sexual reproduction that can produce progenies and molecular markers that behave as genetic markers, that is, that segregate following a Mendelian ratio. When using maps to identify target regions (such as those responsible for the tolerance to abiotic and biotic stresses), important information is generated by association on a trait that is otherwise difficult or costly to measure, or having a high Genotype × Environment interaction (Young, 1996). In this sense, many studies showing genetic maps have been produced for crop species such as rice, soybean, oats, and wheat, identifying many genes and QTLs involved in abiotic and biotic stresses, as well as

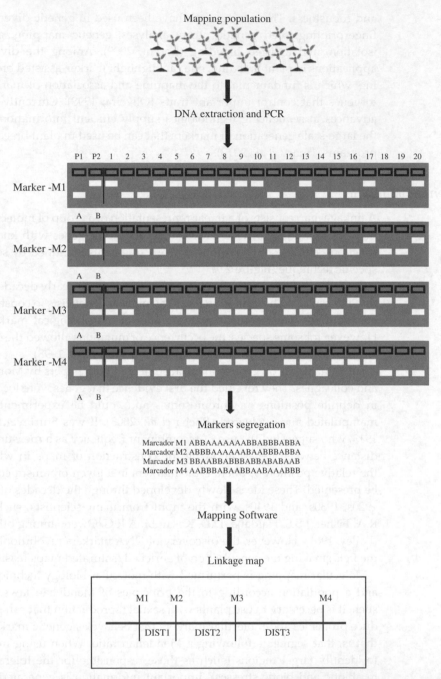

Figure 2.9   Linkage map based on a small population (20 individuals) of recombination inbred lines, based on the segregation of molecular markers. The map is generated by software that analyzes the probabilities of marker orders and distances between them, attaining the best fit (Source: Adapted from Collard et al., 2005). (See color figure in color plate section).

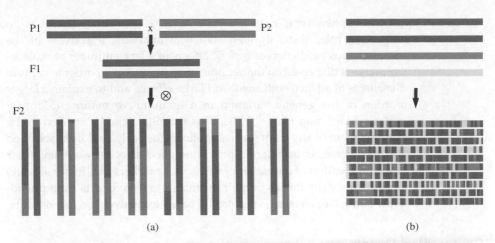

(a)   (b)

Figure 2.10   Comparison of mapping strategies. (a) Biparental mapping is based on populations originating from controlled crosses (the example is showing an F$_2$ population), with few recombination opportunities and large extensions suffering from linkage disequilibrium. (b) Association mapping is performed on populations that have undergone many generations and therefore, many opportunities for recombination. Also, there are small extensions of linkage disequilibrium. The white star represents a cluster of linked genes, showing how rare the occurrence of recombinations is when the loci are very close (Source: Adapted from Zhu et al., 2008). (See color figure in color plate section).

biofortification (Jun, Mian, and Michel, 2012; Jackson et al., 2010; Mason et al., 2010; Norton et al., 2010).

Association Mapping

Distinct from biparental mapping, which is based on populations originating from controlled crosses, association mapping is based on germplasm collections, natural populations and elite genotypes to name just a few (Figure 2.10). This technique considers all the recombinations occurring around the mutation site. In addition, the target regions tend to be smaller than those on conventional (biparental) mapping, generating an increase in both the number of markers needed and in the map resolution (Mackay, Eric, and Julien, 2009; Gupta, Rustgi, and Kulwal 2005). Besides its use on QTL mapping, association mapping can identify SNP mutations, responsible for some specific genotypes (Palaisa et al., 2004).

Genome Wide Selection (GWS)

The large contribution of molecular biology to plant breeding is the use of genomic and transcriptomic information for selecting superior genotypes, in order to reach high selective efficiency, accelerate genetic progress, and reduced costs, in comparison with traditional selection based on phenotypic data (Resende *et al.*, 2008). Thus, GWS has an important role, since, through statistical analyses, it consists of the simultaneous prediction of genetic effects of a large number of molecular markers dispersed on the genome of an organism, in order to capture the effects of all loci, both small and large effects, and to explain a larger portion of the genetic variation in a quantitative manner (Meuwissen, Goddard, and Hayes, 2001). This technique can be applied to the improvement of any plant species. It allows the early and direct selection of target genes, in addition to providing good selective accuracy. After the identification of a number of molecular markers that increase selection accuracy (by means of their estimated genetic effects from phenotypic data), the selection is performed based exclusively on marker data.

Comparative Genomics

Comparative genomics is the study of similarities and differences of the structure and function of the hereditary information among taxa. The necessary analyses are performed by means of molecular biology techniques and computational analyses. In plants, the evolution of small but essential segments in the genome occurs slowly, allowing the recognition of common intragenic regions between species that diverged a long time ago, as well as similar gene arrangements on the chromosomes, showing a synteny among genomes (Moore *et al.*, 1995). A comparison of genomes at the sub-centimorgan level establishes that gene colinearity is maintained (Chen *et al.*, 1997). The variations in the levels of colinearity and synteny are given by diverse factors, such as duplications and chromosomal segmentations, transposable element mobility (transposon and retrotransposon), gene deletion, and local rearrangements (Paterson *et al.*, 2000).

Comparative studies between species have always been of great scientific importance in understanding evolutionary aspects and have been shown to be a powerful tool for investigating genes and genome evolution, and also for improving genome annotation (Lu *et al.*, 2009). The level of gene conservation is an important criterion to determine the extent to which comparative knowledge can be applied between various species. Comparative genomic analyses have shown that the order of genes between species and related plants has remained largely conserved through millions of years of evolution. Therefore, the major goals of comparative genomics studies are to compare the organization of related genomes, make inferences on the basic processes of genome evolution, transfer the information from model to related species, and to integrate the information on the localization of a gene and its expression for a given species (Devos, 2005).

Besides *Arabidopsis thaliana*, a model dicot species, rice is largely used in comparative genome research. Rice has economical and cultural value and is considered a model species because its genome is the smallest among the important cereals. Rice has a genome size of 370 Mbp (million base pairs), small, when compared with maize (2500 Mbp), barley (4900 Mbp), and wheat (16 000 Mbp). The economic and scientific importance of rice lead to the creation of an international consortium called the IRGSP (International Rice Genome Sequence Project) formed by Japan, the United States, China, France, India, Taiwan, Thailand, South Korea, Brazil, and the United Kingdom (IRGSP, 2005).

Another important project that is based on Comparative Genomics is the IOMAP (International Oryza Map Alignment Project). The objective of this project is to explore the genetic potential of wild rice species. The study was first performed by aligning BACs on physical maps and sequencing from wild and cultivated rice species, extending to a wide study on the genus *Oryza* (Wing *et al.*, 2005). The accumulation of comparative genomics results has transformed the grasses (especially the cereals) into a unique genetic system. Therefore the gains obtained from one species can also benefit other less favored cereals, that is, without the need for more technology or investment (Varshney *et al.*, 2006; Bennetzen and Freeling, 1993).

Through the analysis of genetic maps of grasses such as oat, wheat, maize, millets, sorghum, and sugarcane (Devos, 2005; Devos and Gale, 1997; Moore *et al.*, 1995), it is possible to identify the formidable conservation of the order and content of genes among the species.

Structural and Comparative Genomics

Plant breeding has been supporting large advances in food production worldwide. However, some current challenges have to be faced, such as the limit on expanding agricultural areas, climatic changes, and the need for an increase in food quality and quantity. As there is now a large capacity for sequencing data generation, the integration of structural and functional genomics appears as a significant tool for plant breeding research. The identification of homologous regions among related species, provided there is synteny, can be used to find candidate genes in related species. For example, the identification of important genes in orphan species, with a smaller set of information, can be improved by investigating the corresponding region on the related model species and narrowing down the number of candidate genes. This strategy allows faster progress in genetic gains and a better understanding of relationships among species.

Applications of Genomics in Plant Breeding

Plant breeding has stimulated an increase in the yields of the majority of cultivated plants, through the development of cultivars with resistance to biotic and abiotic stresses. However, recent data from agriculture indicate that the rate of these yield increases, for many cultivated species, has decreased since the 1970s. This has been observed for species that represent a large portion of world food production, such as maize, wheat, and rice (Tanksley and McCouch, 1997). To recover the former gains in the yields of the genetic indices or to maintain and even surpass these indices are two of the major challenges for breeding programs over the coming years. In this context, genomics appears to be one of the tools that can contribute to overcoming these challenges, especially on abiotic stress research, which is among the major factors involved in cultivating sub-optimal areas. Therefore, understanding the genetic basis of particular traits will have a major role to play in driving plant breeding forward to overcome those challenges (Costa de Oliveira and Varshney, 2011; Fritsche-Neto and Borém, 2011).

References

Allard, R.W. 1971. Princípios do melhoramento genético das plantas. São Paulo: Edgard Blucher Publishers.

Arabidopsis Genome Initiative. 2000. Analysis of the genome sequence of the flowering plant *Arabidopsis thaliana*. Nature, 408 (6814): 796–815.

Argout, X. Salse, J.; Aury, J.M.; *et al.* 2011. The genome of *Theobroma cacao*. Nature Genetics, 43: 101–108.

Bailey, N.T.J. 1961. Introduction to the mathematical theory of genetic linkage. Oxford: Oxford University Press. 421 pp.

Bennetzen, J.L.; Freeling, M. 1993. Grasses as a single genetic system: genome composition, collinearity and compatibility. Trends in Genetics, 9 (8): 259–261.

Bentley, D.R. 2006. Whole-genome re-sequencing. Current Opinion in Genetics and Development, 16: 545–552.

Borém, A. 2009. Aplicação dos marcadores moleculares no melhoramento. In: Borém, A. and Caixeta, E.T. (eds). Marcadores Moleculares. 2nd edn. Viçosa: UFV Publishers, pp. 95–102.

Borlaug, N.E. 1983. Contributions of conventional plant breeding to food production. Science, 219: 689–693.

Botstein, D.; White, R.L.; Skolnick, M.; Davis, R.W. 1980. Construction of a genetic linkage map in man using restriction fragment length polymorphisms. American Journal of Human Genetics, 32: 314–331.

Chen, M.; SanMiguel, P.; Oliveira, A.C.; *et al.* 1997. Microcolinearity in *sh2*-homologous regions of the maize, rice, and sorghum genomes. Proceedings of the National Academy of Sciences U.S.A., 94 (7): 3431–3435.

Collard, B.C.Y.; Jahufer, M.Z.Z.; Brouwer, J.B.; Pang, E.C.K. 2005. An introduction to markers, quantitative trait loci (QTL) mapping and marker-assisted selection for crop improvement: The basic concepts. Euphytica, 142 (1-2): 169–196.

Connell, C.; Fung, S.; Heiner, C.; *et al*. 1987. Automated DNA-sequence analysis. Biotechniques, 5: 342–348.

Costa De Oliveira, A.; Varshney, R.K. 2011. Introduction to root genomics. In: Costa de Oliveira, A. and Varshney, R.K. (eds). Root Genomics. Berlin: Springer. pp. 1–10.

Devos, K.M. 2005. Updating the 'crop circle'. Current Opinion in Plant Biology. 8 (2): 155–162.

Devos, K.M.; Gale, M.D. 1997. Comparative genetics in the grasses. Plant and Molecular Biology. 35 (1-2): 3–15.

Dressman, D.; Yan, H.; Traverso, G.; *et al*. 2003. Transforming single DNA molecules into fluorescent magnetic particles for detection and enumeration of genetic variations. Proceedings of the National Academy of Sciences U.S.A., 100 (15): 8817–8822.

Food and Agricultural Organization of the United Nations. 2010. The State of Food Insecurity in the World 2010. Rome: FAO.

Fritsche-Neto, R.; Borém, A. 2011. Melhoramento de plantas para condições de estresses abióticos. Visconde do Rio Branco: Suprema Publishers. 250 pp.

Gupta, P.K.; Rustgi, S.; Kulwal, P.L. 2005. Linkage disequilibrium and association studies in higher plants: present status and future prospects. Plant Molecular Biology, 57: 461–485.

Hamilton, J.P.; Buell, C.R. 2012. Advances in plant genome sequencing. Plant Journal, 70: 177–190.

Henry, R.J.; Edwards, M.; Waters, D.L.E.; *et al*. 2012. Application of large scale sequencing to plants. Journal of Bioscience, 37 (5): 829–814.

International Rice Genome Sequencing Project. 2005. The map-based sequence of the rice genome. Nature, 1 (7052): 793–800.

Jaccoud, D.; Peng, K.; Feinstein, D.; Kilian, A. 2001. Diversity arrays: a solid state technology for sequence information independent genotyping. Nucleic Acids Research 29 (4): 1–7.

Jackson, E.W.; Obert, D.E.; Avant, J.B.; *et al*. 2010. Quantitative trait loci in the Ogle/TAM O-301 oat mapping population controlling resistance to *Puccinia coronata* in the field. Phytopathology, 100 (5): 484–492.

Jaillon, O.; Aury, JM.; Noel, B.; *et al*. 2007. The grapevine genome sequence suggests ancestral hexaploidization in major angiosperm phyla. Nature, 449: 463–467.

Jun, T.; Mian, M.A.R.; Michel, A.P. 2012. Genetic mapping revealed two loci for soybean aphid resistance in PI 567301B. Theoretical and Applied Genetics, 124 (1): 13–22.

Kalendar, R.; Grob, T.; Regina, M.; *et al*. 1999. IRAP and REMAP: two new retrotransposon based DNA fingerprinting techniques. Theoretical and Applied Genetics, 98: 704–711.

Lu, F.; Ammiraju, J.S.S.; Sanyal, A.; *et al*. 2009. Comparative sequence analysis of *MONOCULM1*-orthologous regions in 14 *Oryza* genomes. Proceedings of the National Academy of Sciences U.S.A., 106 (6): 2071–2076.

Mackay, T.F.C.; Eric, A.S.; Julien, F. A. 2009. The genetics of quantitative traits: challenges and prospects. Nature Reviews Genetics, 10 (8): 565–577.

Mason, R.; Mondal, S.; Beecher, F.W.; *et al*. 2010. QTL associated with heat susceptibility index in wheat (*Triticum aestivum* L.) under short term reproductive stage heat stress. Euphytica, 174: 423–436.

Maxam, A.M.; Gilbert, W. 1977. A new method for sequencing DNA. Proceedings of the National Academy of Sciences U.S.A., 74(2): 560–564.

Metzker, M.L. 2009. Sequencing technologies – the next generation. Nature Reviews Genetics, 11(1): 31–46.

Meuwissen, T.H.E.; Goddard, M.E.; Hayes, B.J. 2001. Prediction of total genetic value using genome-wide dense marker maps. Genetics, 157: 1819–1829.

Ming, R.; Hou, S.; Feng, Y.; et al. 2008. The draft genome of the transgenic tropical fruit tree papaya (Carica papaya Linnaeus). Nature, 452: 991–996.

Moore, G.; Devos, K.M.; Wang, Z.; Gale, M.D. 1995. Grasses, line up and form a circle. Current Biology, 5(7): 737–739.

Morrell, P.L.; Buckler, E.S.; Ross-Ibarra, J. 2012. Crop genomics: advances and applications. Nature Reviews. Genetics, 13: 85–96.

Norton, G.J.; Deacon, C.M.; Xiong, L.; Huang, S.; et al. 2010. Genetic mapping of the rice ionome in leaves and grain: identification of QTLs for 17 elements including arsenic, cadmium, iron and selenium. Plant and Soil, 329(1): 139–153.

Oliveira, A.C. 1998. Construção de mapas genéticos em plantas. In: Milach, S.C.K. Marcadores moleculares em plantas. Porto Alegre: UFRGS Publisher. 141 pp.

Palaisa, K.; Morgante, M.; Tingey, S.; Rafalski, A. 2004. Long-range patterns of diversity and linkage disequilibrium surrounding the maize Y1 gene are indicative of an asymmetric selective sweep. Proceedings of the National Academy of Sciences U.S.A., 101 (26): 9885–9890.

Paterson, A.H.; Bowers, J.E.; Burow, M.D.; et al. 2000. Comparative genomics of plant chromosomes. Plant Cell, 12(9): 1523–1540.

Paterson, A.H.; Bowers, J.E.; Bruggmann, R. et al. 2009. The Sorghum bicolor genome and the diversification of grasses. Nature, 457: 551–556.

Resende, M.D.V.; Lopes, P.S.; Silva, R.L.; Pires, I.E. 2008. Seleção genômica ampla (GWS) e maximização da eficiência do melhoramento genético. Pesquisa Florestal Brasileira. Colombo, 56: 63–78.

Rocha, R.B.; Pereira, J.F.; Cruz, C.D.; et al. 2003. Mapeamento genético no melhoramento de plantas. Biotecnologia Ciência e Desenvolvimento, 30: 27–32.

Sanger, F.; Nicklen, S.; Coulson, A.R. 1997. DNA sequencing with chain-terminating inhibitors. Proceedings of the National Academy of Sciences U.S.A., 74: 5463–5467.

Service, R.F. 2006. The race for the $1000 genome. Science, 311: 1544–1546.

Shendure, J.; Ji, H. 2008. Next-generation DNA sequencing. Nature Biotechnology, 26: 1135–1145.

Tanksley, S.D.; Mccouch, S.R. 1997. Seed banks and molecular maps: unlocking genetic potential from the wild. Science, 277: 1063–1066.

Thudi, M.; Jackson, S.A.; May, G.D.; Varshney, R.K. 2012. Current state-of-art of sequencing technologies for plant genomics research. Briefings in Functional Genomics, 11(1): 3–11.

Tuskan, G.A.; Difazio, S.; Jansson, S.; et al. 2006. The genome of black cottonwood, Populus trichocarpa (Torr. and Gray). Nature, 313 (5793): 1596–1604.

Varshney, R.K.; Graner, A.; Sorrells, M.E. 2005. Genic microsatellite markers in plants: features and applications. Trends in Biotechnology. 23: 48–55.

Varshney, R.K.; Hoisington, D.A.; Tyagi, A.K. 2006. Advances in cereal genomics and applications in crop breeding. Trends in Biotechnology, 24 (11): 490–499.

Varshney, R.K.; Nayaki, S.N.; May, G.D.; Jackson, S.C. 2009. Next-generation sequencing technologies and their implications for crop breeding. Trends in Biotechnology, 27(9): 522–530.

Vogel, J.P.; Garvin, D.F.; Mockler, T.C.; *et al.* 2010. Genome sequencing and analysis of the model grass *Brachypodium distachyon.* Nature, 463 (7282): 763–768.

Wing, R.A.; Ammiraju, J.S.; Luo, M.; *et al.* 2005. The *Oryza* Map Alignment Project: The golden path to unlocking the genetic potential of wild rice species. Plant Molecular Biology, 59: 53–62.

Xu, Y.; Li, Z.K.Y.; Thomson, M. 2012. Molecular breeding in plants: moving into the mainstream. Molecular Breeding, 29 (4): 831–832.

Young, N.D. 1996. QTL mapping and quantitative disease resistance in plants. Annual Review of Phytopathology, 34(1): 479–501.

Zhu, C.; Gore, M.; Buckler, E.S.; Yu, J. 2008. Status and prospects of association mapping in plants. The Plant Genome, 1 (1): 5–20.

3 Transcriptomics

Carolina Munari Rodrigues, Valéria S. Mafra,
and Marcos Antonio Machado
*Department of Biotechnology, Center for Citriculture Sylvio Moreira,
Agronomical Institute of Campinas, Cordeirópolis, SP, Brazil*

Transcriptomics is the field of molecular biology that studies the transcriptome: the complete set of transcripts in a cell, tissue or organism, which includes the messenger RNA (mRNA) and non-coding RNA (ncRNA) molecules (Morozova *et al.*, 2009). Unlike the genome, which is practically fixed for a given organism (with few exceptions), the transcriptome is more dynamic, and can vary in a particular cell, tissue or organ depending on the developmental stage and in response to external stimuli. Thus, transcriptomics is considered to be the major large-scale platform for studying the biology of an organism. Transcriptome analysis has been widely used in various fields of biology, such as genetics, biochemistry, ecology, and evolution (Kliebestein, 2012). By studying the transcriptome, researchers can determine which sets of genes are turned on or off in a particular condition and quantify the changes in gene expression among different biological contexts. With the advent of high-throughput sequencing technologies, it is also possible to map transcripts onto the genome, obtaining valuable information about gene structure, splicing patterns and other transcriptional modifications (Wang, Z. Gerstein, and Snyder, 2009). As a result, numerous genes, alleles, and splice variants have been identified in various organisms.

In recent decades, numerous methods have been developed to identify and quantify the transcriptome. Such methods have evolved from the candidate gene-based detection of a few transcripts using northern blotting to high-throughput techniques, which detect thousands of transcripts simultaneously, such as microarrays and next-generation sequencing technologies (e.g., RNA-Seq) (Morozova *et al.*, 2009).

This chapter summarizes the main techniques used to study the transcriptome and discusses how different approaches to transcriptome analysis have been applied in plant breeding.

Methods of Studying the Transcriptome

Various techniques and methods for studying a transcriptome can be used. Such methods can be divided into sequencing-based approaches, for example expressed sequence tags (EST) sequencing, serial analysis of gene expression (SAGE), massive parallel signature sequencing (MPSS), RNA-seq, and hybridization-based approaches, such as microarray technology and tiling-arrays. Choosing the best method to survey a particular transcriptome depends on several factors, such as: the complexity of the organism studied; the availability of a sequenced genome or previous knowledge of the transcriptome (e.g., EST libraries); the capacity to store, retrieve, process, and analyze a huge amount of data; and the percentage of transcripts surveyed, which has cost implications.

Construction of cDNA Libraries for EST Sequencing

ESTs are short sequence reads derived from the partial sequencing of complementary DNA (cDNA) sequences. The term "EST" was first proposed by Adams and co-workers in 1991. These researchers constructed a library of cloned cDNAs derived from transcripts of the human brain and generated ESTs by partial sequencing of 600 randomly selected clones, resulting in the discovery of new human genes (Adams *et al.*, 1991). This first report was the starting point for the generation of EST databases for myriad organisms. For decades, the information generated by ESTs was the main resource used to identify novel gene transcripts and assess gene expression levels in a given biological context.

Basically, the construction of a cDNA library and the generation of ESTs involve: the isolation of total RNA and purification of mRNAs; synthesis of cDNA using a reverse transcriptase enzyme; cloning of cDNA fragments and sequencing of randomly selected clones using the Sanger method (Sanger *et al.*, 1977) (Figure 3.1). Both the 3′ and 5′ ends of a cDNA clone could be sequenced, resulting in ESTs ranging in size from 100 to 700 bp (Nagaraj *et al.*, 2006).

After generating ESTs, the next step is EST sequence analysis, which can be divided into pre-processing, clustering/assembly, and EST annotation. During pre-processing, vector adaptors and primer sequences are removed from the ESTs, and low quality and very short ESTs are also discarded from the dataset. Low complexity regions, such as repetitive elements and Simple Sequence Repeats (SSR) should also be detected and masked (replacing nucleotides of these regions with "N"). Poly(A)-tails should be trimmed to retain a few adenines (Nagaraj *et al.*, 2006). Pre-processing results in high-quality ESTs suitable for the next step: the clustering and assembly of ESTs.

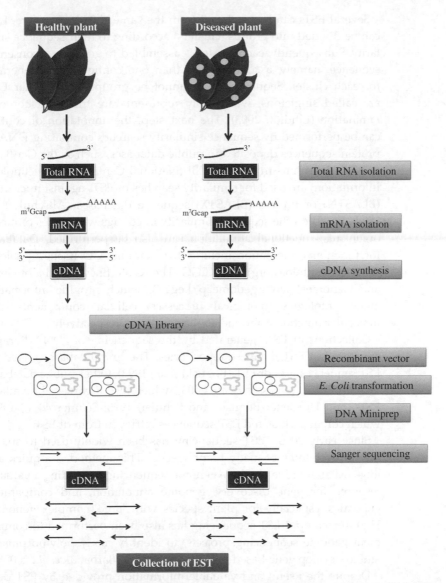

Healthy plant   **Diseased plant**

5'   3'   5'   3'
Total RNA   **Total RNA**   Total RNA isolation

m⁷Gcap   AAAAA   m⁷Gcap   AAAAA
mRNA   **mRNA**   mRNA isolation

5'   3'   5'   3'
3'   5'   3'   5'
cDNA   **cDNA**   cDNA synthesis

cDNA library

Recombinant vector

E. Coli transformation

DNA Miniprep

Sanger sequencing

cDNA   **cDNA**

Collection of EST

Figure 3.1   Overview of cDNA construction and ESTs generation. The process starts with the isolation of mRNA from the total RNA individually isolated from samples of interest (in this example, leaf samples of diseased and healthy/control plants). mRNAs are reverse transcribed to produce complementary DNA (cDNAs) creating libraries of cDNAs cloned into an appropriate vector. After cloning and *E. coli* transformation, individual clones from these libraries are randomly selected and subjected to a single sequencing reaction using universal primers; one or both ends of the cloned fragment (insert) are sequenced. The collection of short fragments (ESTs) generated is then processed using a number of bioinformatics tools. In this example, it is also possible to use the relative abundance of ESTs representing the same gene to compare the level of gene expression between two experimental conditions. (See color figure in color plate section).

Several ESTs can be produced from the same cDNA; therefore, ESTs can be divided into groups (clusters) according to their sequence similarity. Subsequently, each cluster is assembled to generate a consensus sequence, namely, a contig. More than one contig can be generated for each cluster. Sequences that cannot be grouped with other ESTs are called singletons, which may represent rare transcripts or contamination (Lindlöf, 2003). The next step, the annotation of contigs, can be performed by sequence similarity searches comparing DNA or protein sequences deposited in public databases, such as the GenBank. BLAST algorithms from the NCBI (National Center for Biotechnology Information) are used for similarity searches of ESTs against nucleotide (BLASTN) or protein (BLASTX) sequence databases (Altschul *et al.*, 1990). Based on the sequence similarity of contigs with genes of model organisms, functional annotation can also be performed. Databases for functional annotation include, for example, GO (Gene Ontology; www.geneontology.org) and KEGG (The Kyoto Encyclopedia of Genes and Genomes; www.genome.jp/kegg/), which provide information about ontologies (biological processes, cellular components, and molecular functions) and metabolic pathways, respectively.

Collections of ESTs generated by the sequencing of cDNA libraries are stored in EST database repositories. The first database created for EST sequences was dbEST (Boguski *et al.*, 1993). Other public databases include EMBL (Stoesser *et al.*, 2003), which also archives all available ESTs, and UniGene (Boguski and Schuler, 1995; Schuler *et al.*, 1996), which contains clustered EST sequences retrieved from dbEST.

Since early 2000s, EST sequencing has been widely used to survey the transcriptome of many plant species. This method is a quick and low-cost alternative to whole genome sequencing, providing a valuable resource for gene discovery, genome annotation, and comparative genomics, especially for plant species with large, complex genomes. High-throughput EST sequencing has also been performed to complement genome sequencing projects, to identify many polymorphisms, and to develop gene-based molecular markers (Barbazuk *et al.*, 2007).

Despite the useful and valuable information provided by EST data, EST sequencing has a number of limitations. ESTs are sequenced only once; therefore, they are often of low quality and may have a high error rate. These errors result from substitutions, deletions, and insertions of nucleotides when compared with the original mRNA sequence (Lindlöf, 2003). EST libraries are also subject to contamination by sequences generated from genomic DNA, vector DNA, chimeric cDNAs (artifacts produced during ligation and reverse-transcription reaction), mitochondria or ribosomal DNA (Nagaraj *et al.*, 2006). Poorly expressed transcripts are often missed or under-represented within libraries, while highly expressed transcripts are over-represented. If an EST is not represented within an EST library, it does not mean that the corresponding gene is not expressed or absent from the genome (Rudd, 2003). Moreover, EST sequencing by the Sanger method is laborious, time-consuming, and remains expensive for surveying large-scale transcriptomes. Thus, EST data are not suitable for estimating transcript abundance (Morozova *et al.*, 2009).

Suppression Subtractive Hybridization (SSH)

Suppression subtractive hybridization (SSH) is a method used for separating either DNA or cDNA molecules that distinguish two related samples. In particular, the SSH technique can be applied to study transcriptomics, as it is a comparative method that examines the relative abundance of transcripts of a sample of interest (tester) in relation to a control sample (driver) (Luk'ianov *et al.*, 1994; Diatchenko *et al.*, 1996; Gurskaya *et al.*, 1996). SSH is based on the principle of PCR suppression, which allows the amplification of desired sequences and simultaneously suppresses the amplification of undesirable ones. In addition, SSH combines normalization and subtraction during the hybridization step, which removes cDNAs that are common between the tester and driver samples and normalizes (equalizes) cDNAs with different concentrations. Normalization occurs because during hybridization, more abundant cDNA molecules can anneal faster than less abundant cDNA fragments that remain single-stranded (Figure 3.2) (Lukyanov *et al.*, 2007). Finally, the method retains only differentially expressed or variable sequence transcripts that were present in the tester (Desai *et al.*, 2001).

Briefly, the SSH technique starts with RNA isolation from the tester and driver samples, followed by mRNA isolation and cDNA synthesis. Tester and driver cDNA samples are digested with a blunt cutting restriction enzyme, and the resulting tester cDNA fragments are then subdivided in two aliquots. Each aliquot is ligated with a different adapter (adapters 1 and 2), which contain self-complementary ends (Figure 3.2). After two rounds of hybridization, only target double-stranded fragments that are present in the tester and have different ligated adapters can be exponentially amplified. After denaturing and annealing, single-stranded fragments containing the same adapter at both ends form a stem–loop structure and their amplification is suppressed (Figure 3.2). Once a subtracted sample is confirmed to be enriched in cDNAs derived from differentially expressed genes, the fragments can be cloned or subcloned, generating a subtracted cDNA library. Subsequently, selected clones are sequenced and the functional annotation of sequences can be performed using the same procedure described in the EST sequencing method.

SSH technique is a powerful approach for detecting differential gene expression of organisms with no previous information of their genome or transcriptome. The procedure has the advantage of eliminating physical separation of ss- and ds-cDNAs, requiring only one round

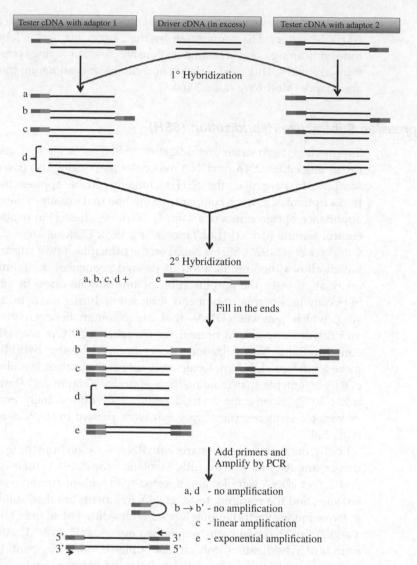

Figure 3.2   Scheme of the suppression subtractive hybridization method. This method includes several steps. Firstly, cDNAs from tester and driver samples are synthesized and digested with a restriction enzyme. Adapters (1 and 2) are ligated to one of two aliquots of tester sample. The first hybridization is performed mixing an excess of driver with both aliquots of the tester. In this step, the subset of tester molecules is normalized: more abundant cDNA fragments anneal faster forming double-stranded (ds) homohybrids (type "b"), while less abundant cDNA fragments remain single-stranded (ss) (type "a"). Also, complementary fragments originating from the same cDNA anneal, resulting in tester-driver double-stranded heterohybrids (type "c"). Secondly, the two samples from the first hybridization are pooled and more freshly denatured driver is added. During the second hybridization, only ss-molecules (type "a") are able to reassociate and form type "b", "c", and "e" hybrids. The least is the tester–tester heterohybrid, having a different adapter at each end. After the second hybridization, end-filling and PCR, "a" and "d" molecules lack primer annealing sites and cannot be amplified; "b" molecules form stem–loop structures that suppress amplification (suppressive effect); "c" molecules are linearly amplified because they have only one primer annealing site. Type "e" molecules are exponentially amplified using primers that anneal at the two different primer annealing sites. Scheme according to the instruction manual (PCR- Select™ cDNA subtraction kit, Clontech). (See color figure in color plate section).

of subtractive hybridization. The two hybridization steps lead to an efficient normalization of cDNA concentrations, achieving over 1000-fold enrichment for differentially expressed cDNAs, including rare transcripts. However, in practice, the level of enrichment depends on several factors, such as the concentration of tester relative to the driver sample, the number of differentially expressed genes, the nature and complexity of the samples used, and the time of hybridization (Lukyanov *et al.*, 2007). Also, SSH techniques can generate many false positives, which are undesirable (background) clones representing non-differentially expressed genes. In some cases, the number of false positives can exceed the number of target clones in the subtracted library. To overcome this problem, a modified method, known as "mirror orientation selection" can be used to decrease the number of false positive clones (Rebrikov *et al.*, 2000).

While SSH is not as widely used as microarrays, some studies have reported the use of SSH to survey differential expression in plants. For example, using a modified protocol of SSH, Yao and co-workers identified differentially expressed transcripts in the leaf and root between wheat hybrids and their parents. These transcripts correlated with the observed heterosis in both aerial growth and root related traits (Yao *et al.*, 2005). In another study, SSH was carried out on root cDNA from bulked boron tolerant and intolerant double haploid barley lines grown under moderate boron stress. Among the 111 clones identified within the subtracted library, nine of them were mapped to a previously reported boron tolerance quantitative trait locus (QTL) (Hassan *et al.*, 2010).

Serial Analysis of Gene Expression (SAGE)

SAGE was the first sequencing-based method to quantify the abundance of thousands of transcripts simultaneously (Morozova *et al.*, 2009). It is based on the principle that a short DNA sequence (tag) of 9–11 bp, derived from a defined location within one transcript, contains enough information to uniquely identify that transcript, if the position of the sequence within the transcript is known. By counting the tags, SAGE provides an estimate of the abundance of the transcripts (Vesculescu *et al.*, 1995). Briefly, the method involves the generation of a library of clones containing concatenated short sequence tags from a population of mRNA transcripts. These clones are sequenced using standard Sanger sequencing; tag sequences are matched to reference sequences in other databases to identify the transcripts, and the tags are counted to estimate the relative abundance of the corresponding transcripts. The frequency of these transcripts in two or more SAGE libraries can be compared to distinguish differences in gene expression in the respective samples (Madden *et al.*, 2000) (Figure 3.3).

SAGE offers some advantages compared with other methods, such as the EST sequencing and microarrays. In SAGE, instead of a clone each

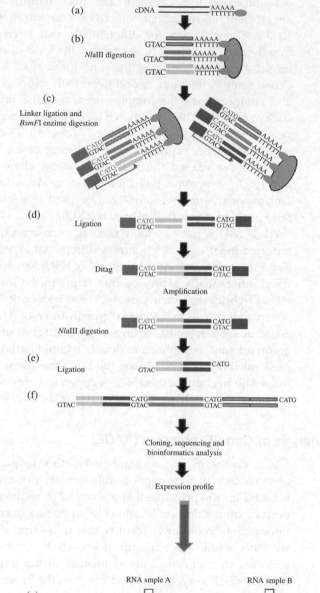

(a) cDNA

(b) NlaIII digestion

(c) Linker ligation and *BsmF*I enzyme digestion

(d) Ligation

Ditag

Amplification

NlaIII digestion

(e) Ligation

(f)

Cloning, sequencing and bioinformatics analysis

Expression profile

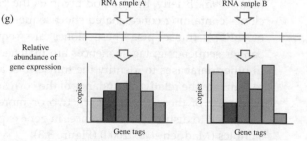

(g) RNA smple A   RNA smple B

Relative abundance of gene expression

copies   copies

Gene tags   Gene tags

cDNA fragment representing only one transcript in a plasmid vector, multiple tags derived from different transcripts are present in a unique plasmid vector. This strategy improves the throughput of data generated per sequencing run and reduces costs, compared with EST sequencing (Vega-Sánchez *et al.*, 2007). Unlike the microarray technology, which relies on previously identified sequences, SAGE has the ability to discover novel transcripts and to detect poorly expressed transcripts. By counting tags, SAGE obtains a direct measure of transcript abundance. The data obtained can also be easily compared between multiple samples and across different experiments (Morozova *et al.*, 2009).

The major drawback of SAGE is the difficulty of identifying and annotating tags unambiguously, because short tags often match with multiple genes with similar coding sequences (Morozova *et al.*, 2009). In addition, reliable annotation depends on the existence of comprehensive EST or genomic databases to be used as reference (Madden *et al.*, 2000). To overcome some limitations of SAGE, modifications to the original methodology were proposed, such as LongSAGE (Saha *et al.*, 2002; Wei *et al.*, 2004) and SuperSAGE (Matsumura *et al.*, 2003), which result in tags with 21 and 26 bp, respectively, and DeepSAGE, which uses LongSAGE-derived ditags and replaces the Sanger sequencing with 454 pyrosequencing (Nielsen *et al.*, 2006).

In the past, SAGE has been applied extensively in medical research to profile the transcriptome of a range of human diseases, including cancer (Zhang *et al.*, 1997; Nacht *et al.*, 1999). However, since the first report of SAGE in plant research (Matsumura *et al.*, 1999), this method has not been widely used by plant biologists compared with other methods of studying large-scale transcriptomes, such as EST sequencing, microarrays, and more recently, RNA-seq.

Figure 3.3 Schematic representation of serial analysis of gene expression (SAGE). (a) cDNA synthesis using reverse transcriptase and oligo-dT primers attached to magnetic beads. (b) cDNAs are digested with the restriction endonuclease Nla III. (c) Adapters are ligated to the digested cDNA. These adapters have a recognition site for the enzyme BsmF I, which cleaves at a fixed distance downstream from its recognition site. Cleavage releases the sequences from the magnetic beads. (d) Each released cDNA tag is connected to another forming a ditag. (e) Ditags are amplified by PCR using primers that anneal to the adapters. Subsequently, the amplified fragments are cleaved with Nla III to release the adapters. (f) Ditags are ligated to form a concatemer, which is then cloned into vectors and sequenced, generating the SAGE library. (g) The number of tags found within the same sequence permits us to deduce the relative abundance of gene expression. (Source: Diagram from Garnis *et al.*, 2004). (See color figure in color plate section).

Microarrays

Since their conception, microarrays have revolutionized the study of large-scale transcriptomes in different organisms and biological contexts (Schena *et al.*, 1995). This technology consists of an array of single-stranded DNA molecules, called probes (fragments of genomic DNA, cDNA or oligonucleotides) chemically linked onto a solid surface, usually glass slides, which are called chips.

Briefly, to compare different samples, mRNA is isolated from two populations and used as templates for labeled cDNA synthesis with Cye3- and Cye5-dUTP. These labeled cDNAs are hybridized against the probes immobilized on the chip. After this step, the attached fluorophores are excited by a laser to produce specific spectrums that are captured by a scanner. Using software, the fluorescence intensity is proportionally related to the level of gene expression, which enables the generation of the relative abundance of expression between samples (Brown, 1995) (Figure 3.4).

In a single DNA microarray it is possible to analyze the differential expression of about 10 000–40 000 targets per chip (Mockler *et al.*, 2005). Thus, this technique allows the simultaneous analysis of gene expression levels of thousands of genes.

Another approach to this technique is genomic analysis, using high-density oligonucleotide-based whole-genome microarrays. This platform permits the analysis of alternative splicing, characterization of the methylome, polymorphism discovery and genotyping, comparative genome hybridization, and genome resequencing (Mockler *et al.*, 2005).

Although microarrays have the ability to evaluate a large number of transcripts simultaneously, the greatest disadvantage of this method is that they detect only the transcripts of genes that have been previously characterized. Furthermore, this technique is an indirect method to measure the transcripts abundance, as it is inferred by the hybridization signal and is not a direct measure of the transcripts, such as, for example, RNA-seq, which which is explained below (Morozova *et al.*, 2009).

RNA-seq

RNA-seq is a high-throughput sequencing of cDNA, and is based on the direct sequencing of transcripts. This technique is more dynamic, reproducible, and provides a better estimate of the absolute expression levels (Nagalakshmi *et al.*, 2008; Fu *et al.*, 2009). These are the main advantages of RNA-seq compared with microarrays. Another advantage is that the analysis of RNA-seq allows us to identify isoforms of a gene, which are not easily detected using microarrays (Wang *et al.*, 2009).

RNA-seq uses next generation sequencing (NGS) methods to sequence cDNA from RNA of biological samples, producing millions of short

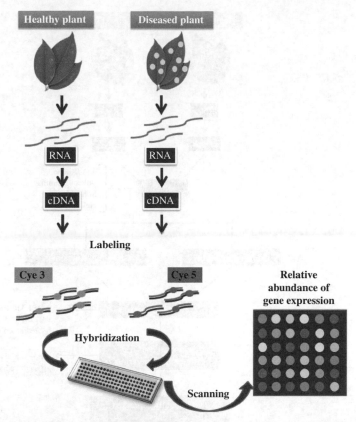

Figure 3.4   Analysis of gene expression using a DNA microarray. Total RNA is isolated from two different samples (healthy and diseased plant), purified, and used as templates for the synthesis of cDNAs, which are labeled with different fluorescent (Cye 3 or Cye 5) dyes for each condition. The labeled cDNAs are mixed and hybridized against the probes (DNA/cDNA known) immobilized on the microarray chip. Laser excitation of the fluorophores produces an emission with a specific spectrum, which is captured by a scanner and analyzed by software. The resulting color intensity is associated with the emitted fluorescence from each sample hybridized, which is proportional to the level of gene expression. (See color figure in color plate section).

reads, whose sizes depend on the platform used (Shendure and Ji, 2008) (Figure 3.5).

Usually, reads are mapped against a reference genome and the number of reads mapped to a region of interest, such as a gene, is used to measure the relative abundance of its expression. Moreover, the programs that perform assembly using the reference genome can identify isoforms from transcripts (Oshlack *et al.*, 2010). However, RNA-seq also allows assembly of transcripts without the use of a reference genome (*de novo* assembly), which makes this technique more advantageous compared with other methods for large-scale analysis of gene expression (Grabherr *et al.*, 2011) (Figure 3.5 and Table 3.1).

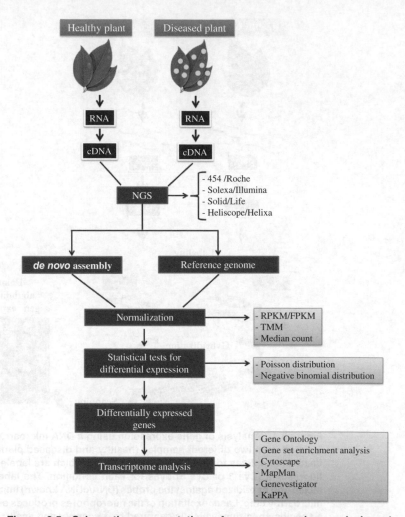

Figure 3.5 Schematic representation of gene expression analysis using RNA-seq. RNA is isolated from two contrasting samples (healthy and diseased plant), and used for the synthesis of cDNAs. These are sequenced using one of the NGS methods, generating millions of reads for each sample. The reads are mapped against a reference genome or *de novo* assembly is performed. After this step, the data are normalized, and then statistical tests are performed to identify differentially expressed genes. Certain programs can infer the biological functions of these genes. (See color figure in color plate section).

After the assembly of the transcripts, the next step is the normalization of the data. An example of a normalization method is performed by Cufflinks, which uses data processed in reads per kilobase of exon model per million mapped reads (RPKM) or fragments per kilobase per million mapped reads (FPKM), when analyzing paired-end sequencing data (Trapnell *et al.*, 2010). These transformations normalize the counts with different library sizes and lengths of the transcripts; a long transcript

Table 3.1   Summary of the main methods and programs for analysis of differential expression using RNA-seq.

Type of analysis	Method	Program	References
Assembly	Assembly of reads using the reference genome/ identification of splice variants	TopHat	Trapnell et al., 2009
		SpliceMap	Au et al., 2010
		ALEXA-seq	Griffith et al., 2010
	Assembly of reads without the use of reference genome	Velvet	Zerbino and Birney, 2008
		Multiple-k	Surget-Groba and Montoya-Burgos, 2010
		Rnnotator	Martin et al., 2010
		Trans-ABySS	Robertson et al., 2010
		Oases	http://www.ebi.ac.uk /zerbino/oases/
		Trinity	Grabherr et al., 2011
Normalization/ differential expression	RPKM	Cufflinks/ Cuffdiff	Trapnell et al., 2010
	TMM	EdgeR	Robinson et al., 2010
	Median count ratio	DESeq	Anders and Huber, 2010

is expected to produce more reads than a short transcript with the same expression level (Soneson and Delorenzi, 2013). However, other programs use different normalization methods (Table 3.1). Some programs calculate the differential expression between two experimental conditions, such as Cufflinks/Cuffdiff and analysis tools implemented in EdgeR and DESeq (Table 3.1).

The analyses of gene expression using RNA-seq are relatively recent and there is still no consensus on the best methods of assembly, normalization, and statistics to calculate the differential expression. Existing methods are still being optimized simultaneously with the development of new methods. Thus, the most appropriate methods of analysis should be carefully selected taking into account each experimental condition.

All high-throughput transcriptome analyzes, such as EST libraries, microarrays, and RNA-seq, generate a list with hundreds or even thousands of differentially expressed genes. Meaningful biological interpretation from a list of differentially expressed genes is a major challenge in the study of the transcriptome. To address this question, several bioinformatics tools have been developed. Among these tools are the functional classification of genes based on Gene Ontology (Harris *et al.*, 2004), and gene set enrichment analysis (GSEA), which relates the terms of GO to identify the most representative ontologies in the list of differentially expressed genes (Alexa *et al.*, 2006). Furthermore, other tools have been developed to integrate gene expression data onto gene regulatory networks or biological pathways, providing a visual representation of the graph and integrated data, such as Cytoscape (Shannon *et al.*, 2003), MapMan (Thimm *et al.*, 2004), GENEVESTIGATOR (Zimmermann *et al.*, 2004) and KaPPA (Tokimatsu *et al.*, 2005).

Finally, after the identification of important genes and pathways regulating the biological context under study, the next step of a global gene expression analysis involves the selection of a set of genes to validate the expression pattern using another method to assess the expression level. This step is generally performed using real-time PCR, which is a more sensitive, specific, and robust method compared with methods used for large-scale profiling of the transcriptome, such as microarrays and RNA-seq (Wong and Medrano, 2005).

Applications of Transcriptomics Approaches for Crop Breeding

Since early 2000s, most plant biology studies focused on the identification of single genes within a biological context. With the advent of large-scale methods to survey the transcriptome, plant scientists could evaluate how thousands of genes work together and study the molecular basis of complex plant processes (Morozova *et al.*, 2009).

Among the different methods for studying the transcriptome, large-scale sequencing of ESTs was the first strategy used for gene discovery and expression profiling in plants. In model plant species, such as *Arabidopsis*, poplar, and rice, extensive EST libraries have been sequenced and annotated to provide transcriptome databases and used as references for comparison with non-model plants (Kumpatla *et al.*, 2012). Over a million ESTs have been deposited in the NCBI database and as of January 2013, there were >74 million ESTs in dbEST (http://www.ncbi.nlm.nih.gov/dbEST/), including ESTs for a number of model and non-model plant species (Table 3.2).

Several EST sequencing projects and EST databases were created for crop species such as maize, sugarcane, and cotton (Table 3.3). Other databases, such as HarvEST (http://harvest.ucr.edu/) and PlantGDB (http://plantgdb.org/prj/) were also developed to provide collections of ESTs derived from various plant species and several bioinformatics tools for EST mining (Duvick *et al.*, 2008).

Table 3.2   Number of ESTs for model and crop plants deposited in NCBI (January 2013).

Plant species[a]	No. of ESTs (dbEST)
Zea mays (maize)	2 019 137
Arabidopsis thaliana (thale cress)	1 529 700
Glycine max (soybean)	1 461 722
Triticum aestivum (wheat)	1 286 372
Oryza sativa (rice)	1 253 557
Hordeum vulgare (barley)	501 838
Vitis vinifera (grapevine)	446 664
Pinus taeda (loblolly pine)	328 662
Malus x Domestica (apple)	325 02
Gossypium hirsutum (cotton)	297 522
Solanum lycopersicum (tomato)	297 142
Citrus sinensis (sweet orange)	214 598
Sorghum bicolor (sorghum)	209 835
Coffea arabica (coffee)	174 275
Theobroma cacao (cocoa)	159 996
Saccharum hybrid cultivar (sugarcane)	135 534
Populus trichocarpa (populus)	89 943
Eucalyptus calmadulensis (eucalyptus)	58 584

[a]Only the most representative plant species are shown.

Table 3.3   Examples of EST sequencing projects and databases for crop species.

Organism	EST project sequencing	Homepage
Apple	Apple EST Project	http://titan.biotec.uiuc.edu/apple/
Citrus	Citrus EST Project (CitEST)	http://biotecnologia .centrodecitricultura.br
Coffee	Brazilian Coffee Genome Project	http://www.lge.ibi.unicamp .br/cafe/
Eucalyptus	Brazilian Eucalyptus Genome Sequence Project Consortium (FORESTs)	https://forests.esalq.usp.br
Maize	Maize Gene Discovery Progress (MGDP)[a]	http://www.maizegdb.org /documentation/mgdp/
Soybean	Soybean EST Project (SoyBase)	http://www.soybase.org/
Sugarcane	Sugarcane EST Genome Project (SUCEST)[b]	http://sucest-fun.org/index.php /projects/sucest
Tomato[c]	SOL Genomics Network (SGN)	http://solgenomics.net/
Wheat	International Triticeae EST Consortium (ITEC)	http://wheat.pw.usda.gov /genome/group/pool/

[a]Since 2003, MGDP has been incorporated into the Maize Genome Database (MaizeGDB; http://www.maizegdb.org/documentation/mgdp/team.php).
[b]SUCEST has been incorporated into the SUCEST-Fun database of the Sugarcane Genome Project (http://sugarcanegenome.org/).
[c]SGN also includes datasets from potato, eggplant and pepper.

Numerous transcripts in different biological contexts have been identified using ESTs. For example, the comparative analysis of two EST libraries generated from infected versus non-infected citrus plants allowed the identification of several citrus transcripts that respond to infection by different pathogens, including *Citrus leprosis* virus (Freitas-Astúa *et al.*, 2007), *Xylella fastidiosa* (De Souza *et al.*, 2007) and *Citrus tristeza* virus (Cristofani-Yaly *et al.*, 2007).

With the progress of large-scale EST analysis in various plant species, accurate mining of these sequences has enabled the identification and mapping of numerous molecular markers. Examples of molecular markers derived from ESTs include EST-SSRs from maize and other cereals (Kantety *et al.*, 2002; Sharapova *et al.*, 2002), citrus (Palmieri *et al.*, 2007), or sugarcane (Vettore *et al.*, 2003) and single nucleotide polymorphism (SNP) markers from tomato (Yamamoto *et al.*, 2005), coffee (Vidal *et al.*, 2010), or eucalyptus (Grattapaglia *et al.*, 2011).

Microarrays have become the favorite transcriptomic approach for many plant biologists. The most extensive microarray studies have been conducted in Arabidopsis thaliana, using the Affymetrix ATH1 array platform (Redman *et al.*, 2004). Large sets of *Arabidopsis* microarray data have been generated and integrated into databases for microarray data mining, such as GENEVESTIGATOR (Zimmermann *et al.*, 2004). Other bioinformatics resources and analysis tools have also been developed for *Arabidopsis* and other model plants (Schmid *et al.*, 2005; Benedito *et al.*, 2008). For example, the database RiceXPro provides a platform for monitoring gene expression of the rice, covering a wide range of tissues, organs, and developmental stages under natural and field conditions (Sato *et al.*, 2011).

With the increase in the number of ESTs and the availability of genomic sequences, Affymetrix designed and released arrays for various crop species, including the Affymetrix GeneChip® array for citrus, *Medicago* spp., cotton, barley, maize, poplar, rice, soybean, tomato, grape, and wheat (http://www.affymetrix.com). Another company, NimbleGen/Agilent has also designed and commercialized arrays for crops (http://www.agilent.com).

Microarrays have a wide range of applications in plant research and plant breeding. To dissect the molecular mechanisms related to biotic stresses, for example, microarrays and, more recently, RNA-seq have been applied to profile thousands of transcripts in tolerant and susceptible citrus genoytpes in response to the infection with different pathogens, such as the bacteria *Candidatus Liberibacter* spp., *Xylella fastidiosa* and *Xanthomonas* spp., and the oomycete *Phytophthora parasitica* (Mafra *et al.*, 2013; Rodrigues *et al.*, 2013; Boava *et al.*, 2012; Albrecht and Bowman, 2012; Cernadas *et al.*, 2008). Intensive research is being carried out to understand the tolerance/susceptibility mechanisms in citrus. The target genes and molecular markers identified from transcriptome

Table 3.4 Summary of recent studies of microarray and RNA-seq in major crops species.

Crops	Goal of study	Method	Reference
Sugarcane	Comparative gene expression profile between sugarcane populations with different sugar content	Microarray	Felix et al., 2009
Maize	Dissecting grain yield pathways and grain dry matter content of maize	Microarray	Fu et al., 2010
Grape	Integrated transcriptome atlas during grape maturation	Microarray	Fasoli et al., 2012
Wheat	Dissecting molecular mechanisms underlying salinity tolerance of a wheat introgression line	Microarray	Liu et al., 2012
Cotton	Dissecting molecular mechanisms related to cotton fiber elongation	Microarray	Gilbert et al., 2013
Soybean	Integrated transcriptome atlas covering different tissues, stages of development and conditions of soybean	RNA-seq	Libault et al., 2010
Rice	Dissecting regulatory pathways involved with grain quality of rice	RNA-seq	Venu et al., 2011
Sorghum	Dissecting molecular mechanisms underlying osmotic stress in sorghum	RNA-seq	Dugas et al., 2011
Citrus	Dissecting molecular mechanisms related to tolerance and susceptibility of citrus against Xylella fastidiosa	RNA-seq	Rodrigues et al., 2013
Tomato	Comparative transcriptome between domesticated and wild tomato	RNAseq	Koenig et al., 2013
Eucalyptus	Dissecting genes and pathways underlying wood formation in three Eucalyptus species	RNA-seq	Salazar et al., 2013

studies have been used in citrus breeding and to produce genetically engineered commercial citrus cultivars to improve their tolerance against these pathogens. Other applications of microarrays to unravel the molecular basis of complex traits are illustrated in Table 3.4.

In traditional QTL analysis, linkage maps lead to the detection of genomic regions that are associated with phenotypic variations within a population (Kumpatla et al., 2012). Microarrays and RNA-seq have been exploited to identify expressed QTLs (eQTLs) using an approach termed "Genetical Genomics" (Jansen and Nap, 2001). The approach combines the genotyping of a segregating population (i.e., genetics) and the large-scale gene expression profiling (i.e., genomics) to identify genetic regulatory loci and genes that explain the observed variation (Joosen *et al.*, 2009). Genetical genomics has been applied, for example, to map eQTLs related to the differentiation of xylem in different genetic backgrounds of Eucalyptus, using gene expression profiles obtained from microarrays. Thousands of transcripts were assessed and 821 of them were mapped to a single eQTL, which was found to explain up to 70% of the observed transcript-level variation (Kirst *et al.*, 2005).

Microarrays can also be applied to understand plant adaptation and domestication (Koenig *et al.*, 2013; Swanson-Wagner *et al.*, 2012). A recent study reported the use of microarrays to examine how domestication has reshaped the transcriptome of maize. By large-scale expression profiling of 18 242 genes for 38 maize and 24 teosinte genotypes, these workers found that significant changes in the expression level between these genotypes correlate with certain genes previously identified as targets of selection during domestication (Swanson-Wagner *et al.*, 2012).

Many studies using microarrays have been published to date; however, as mentioned earlier, because microarrays rely on prior information about EST or genomic sequences, their use has been limited to model plants and well-studied crop species. In recent years, RNA-seq has emerged as an alternative to microarrays for large-scale surveys of the transcriptome. RNA-seq has the potential to provide comprehensive transcriptomic resources for model plants and well-studied crop species where a reference genome is available, but also for less-studied species with no sequenced genome and/or little information about transcript sequences (Varshney *et al.*, 2010). For example, RNA-seq has been used to generate large collections of short transcript reads for several "orphan crops" such as sweet potato, chickpea, and pearl millet (Varshney *et al.*, 2010). For crop species with complex genomes, such as coffee, wheat, and sugarcane, RNA-seq has been used successfully as an alternative to whole genome sequencing, unraveling genes and regulatory regions and providing a myriad of molecular markers (Pérez-de-Castro *et al.*, 2012).

RNA-seq has a wide range of applications. It has been used for several crops with different breeding objectives for transcript discovery and profiling, resulting in the identification of important genes and pathways regulating traits of interest (Table 3.4).

Another interesting application of RNA-seq for plant breeders is the discovery of SNPs. For *Eucalyptus*, for example, about 23 000 SNPs were identified using 454 technology to sequence and assemble 148 Mbp of ESTs (Novaes *et al.*, 2008). Similar efforts were performed in *Brassica* and wheat (Trick *et al.*, 2009, 2012). RNA-seq can also be used to discover and detect noncoding RNA (Morozova *et al.*, 2009). MicroRNAs (miRNA) are small noncoding RNAs that have important roles in regulating several processes in plants by negatively affecting gene expression post-transcriptionally. To understand the role of these small RNAs in response to different conditions and stimuli, high-throughput RNA-seq has been used to identify and profile miRNAs, for example in *Medicago*, rice, and barley in response to drought, general abiotic stresses and during the early stages of seed development, respectively (Wang *et al.*, 2011; Barrera-Figueroa *et al.*, 2012; Curaba *et al.*, 2012).

Conclusions and Future Prospects

The identification of genes affecting economically important traits provides the basis for new progress in the genetic improvement of crop species, complementing traditional breeding methods. Over recent decades, many transcriptomic resources have been generated for model and crop species, and the increase in the number of studies that have adopted the microarrays as the predominant large-scale method of surveying the transcriptome is impressive. However, the generation of a comprehensive transcriptome requires accurate identification and quantification of transcripts using unbiased, highly efficient, and cost-effective methods. In this sense, RNA-seq has emerged as the most powerful transcriptome method, combining all these requirements. However, as an emerging technology, several improvements are being made, such as better storage capacity, faster processing, and better analytical algorithms and tools for downstream analyses and visualization. Finally, integration of transcriptomic and other "omics" data using more sophisticated approaches, such as systems biology, has the potential to dissect the molecular, biochemical, physiological, and evolutionary basis of complex traits. This approach is the great promise of "genomics-assisted breeding" in the 21st century.

Acknowledgements

We thank CNPq and FAPESP for financial support through the National Institute of Science and Technology of Genomics for Citrus Improvement (INCT-Citrus; Proc. 573848/2008-4).

References

Adams, M.D.; Kelley, J M.; Gocayne, J D.; *et* al. 1991. Complementary DNA sequencing: expressed sequence tags and human genome project. Science, 252 (5013): 1651–1656.

Albrecht, U.; Bowman, K.D. 2012. Transcriptional response of susceptible and tolerant citrus to infection with *Candidatus Liberibacter asiaticus*. Plant Science, 185: 118–130.

Alexa, A.; Rahnenführer, J.; Lengauer, T. 2006. Improved scoring of functional groups from gene expression data by decorrelating GO graph structure. Bioinformatics, 22 (13): 1600–1607.

Altschul, S.F.; Gish, W.; Miller, W.; *et al.* 1990. Basic local alignment search tool. Journal of Molecular Biology, 215 (3): 403–410.

Anders, A.; Huber, W. 2010. Differential expression analysis for sequence count data. Genome Biology, 11 (10): R106.

Au, K.F.; Jiang, H.; Lin, L.; Xing, Y.; Wong, W.H. 2010. Detection of splice junctions from paired-end RNA-seq data by SpliceMap. Nucleic Acids Research, 38 (14): 4570–4578.

Barbazuk W.B.; Emrich, S.J.; Chen, H.D.; *et al.* 2007. SNP discovery via 454 transcriptome sequencing. The Plant Journal, 51 (5): 910–918.

Barrera-Figueroa, B.; Gao, L.; Wu, Z.; *et al.* 2012. High throughput sequencing reveals novel and abiotic stress-regulated microRNAs in the influorescences of rice. BMC Plant Biology, 12: 132.

Benedito, V.A.; Torres-Jerez, I.; Murray, J.D.; *et al.* 2008. A gene expression atlas of the model legume *Medicago truncatula*. The Plant Journal, 55 (3): 504–513.

Boava, L.; Cristofani-Yaly, M.; Mafra, V.; *et al.* 2011. Global gene expression of *Poncirus trifoliata*, *Citrus sunki* and their hybrids under infection of *Phytophthora parasitica*. BMC Genomics, 12 (1): 39.

Boguski, M.S.; Lowe, T.M.; Tolstoshev, C.M. 1993. dbEST – database for 'expressed sequence tags'. Nature Genetics, 4: 332–333.

Boguski, M.S.; Schuler, G.D. 1995. ESTablishing a human transcript map. Nature Genetics, 10 (4): 369–371.

Brown, O.P. 1995. Quantitative monitoring of gene expression patterns with a complementary DNA microarray. Science, 270: 467–470.

Cernadas, R.A.; Camillo, L.R.; Benedetti, C.E. 2008. Transcriptional analysis of the sweet orange interaction with the citrus canker pathogens *Xanthomonas axonopodis*pv. *citri* and *Xanthomonas axonopodis*pv. *aurantifolii*. Molecular Plant Pathology, 9 (5): 609–631.

Cristofani-Yaly, M.; Berger, I.J.; Targon, M.L.P.N.; *et al.* 2007. Differential expression of genes identified from *Poncirus trifoliata* tissue inoculated with CTV through EST analysis and *in silico* hybridization. Genetics and Molecular Biology, 30 (3): 972–979.

Curaba, J.; Spriggs, A.; Taylor, J.; *et al.* 2012. miRNA regulation in the early development of barley seed. BMC Plant Biology, 12 (1): 120.

De Souza, A.A.; Takita, M.A.; Coletta-Filho, H.D.; *et al.* 2007. Comparative analysis of differentially expressed sequence tags of sweet orange and mandarin infected with *Xylella fastidiosa*. Genetics and Molecular Biology, 30 (3): 965–971.

Desai, S.; Hili, J.; Trelogan, S.; *et al.* 2001. Identification of differentially expressed genes by suppression subtractive hybridization. In: Stephen, H. and Levesey, R. (eds). Functional Genomics: A Practical Approach. Oxford: Oxford, University Press, pp. 45–80.

Diatchenko, L.; Lau, Y.F.; Campbell, A.P.; *et al.* 1996. Suppression subtractive hybridization: a method for generating differentially regulated or tissue-specific cDNA probes and libraries. Proceedings of the National Academy of Sciences U.S.A., 93 (12): 6025–6030.

Dugas, D.V.; Monaco, M.K.; Olson, A.; *et al.* 2011. Functional annotation of the transcriptome of Sorghum bicolor in response to osmotic stress and abscisic acid. BMC Genomics, 12 (1): 514.

Duvick, J.; Fu, A.; Muppirala, U.; *et al.* 2008. PlantGDB: a resource for comparative plant genomics. Nucleic Acids Research, 36: D959–D965.

Fasoli, M.; Dal Santo, S.; Zenoni, S.; *et al.* 2012. The grapevine expression atlas reveals a deep transcriptome shift driving the entire plant into a maturation program. The Plant Cell Online, 24 (9): 3489–3505.

Felix, J.M.; Papini-Terzi, F.S.; Rocha, F.R.; *et al.* 2009. Expression profile of signal transduction components in a sugarcane population segregating for sugar content. Tropical Plant Biology, 2: 98–109.

Freitas-Astúa, J.; Bastianel, M.; Locali-Fabris, E.C.; *et al.* 2007. Differentially expressed stress-related genes in the compatible citrus-*Citrus leprosis* virus interaction. Genetics and Molecular Biology, 30: 980–990.

Fu, J.; Thiemann, A.; Schrag, T.; *et al.* 2010. Dissecting grain yield pathways and their interactions with grain dry matter content by a two-step correlation approach with maize seedling transcriptome. BMC Plant Biology, 10 (1): 63.

Fu, X.; Fu, N.; Guo, S.; *et al.* 2009. Estimating accuracy of RNA-Seq and microarrays with proteomics. BMC Genomics, 10 (1): 161.

Garnis, C.; Buys, T.P.; Lam, W.L. 2004. Genetic alteration and gene expression modulation during cancer progression. Molecular Cancer, 3 (9): 1–23.

Gilbert, M.K.; Turley, R.B.; Kim, H. J.; *et al.* 2013. Transcript profiling by microarray and marker analysis of the short cotton (*Gossypium hirsutum L.*) fiber mutant Ligon lintless-1 (Li1). BMC Genomics, 14 (1): 403.

Grabherr, M.G.; Haas, B.J.; Yassour, M.; *et al.* 2011. Full-length transcriptome assembly from RNA-Seq data without a reference genome. Nature Biotechnology, 29 (7): 644–652.

Grattapaglia, D.; Silva-Junior, O.B.; Kirst, M.; *et al.* 2011. High-throughput SNP genotyping in the highly heterozygous genome of Eucalyptus: assay success, polymorphism and transferability across species. BMC Plant Biology, 11 (1): 65.

Griffith, M.; Griffith, O.L.; Mwenifumbo, J.; *et al.* 2010. Alternative expression analysis by RNA sequencing. Nature Methods, 7 (10): 843–847.

Gurskaya, N.G.; Diatchenko, L.; Chenchik, A.; *et al.* 1996. Equalizing cDNA subtraction based on selective suppression of polymerase chain reaction: cloning of Jurkat cell transcripts induced by phytohemaglutinin and phorbol 12-myristate 13-acetate. Annual Biochemistry, 240: 90–97.

Harris, M.A.; Clark, J.; Ireland, A.; *et al.* 2004. The Gene Ontology (GO) database and informatics resource. Nucleic Acids Research, 32 (Database issue), D258.

Hassan, M.; Oldach, K.; Baumann, U.; *et al.* 2010. Genes mapping to boron tolerance QTL in barley identified by suppression subtractive hybridization. Plant, Cell & Environment, 33 (2): 188–198.

Jansen, R.C.; Nap, J.P. 2001. Genetical genomics: the added value from segregation. Trends in Genetics, 17 (7): 388–391.

Joosen, R.V.L.; Ligterink, W.; Hilhorst, H.W.M.; Keurentjes, J.J.B. 2009. Advances in genetical genomics of plants. Current Genomics, 10 (8): 540.

Kantety, R.V.; La Rota, M.; Matthews, D.E.; Sorrells, M.E. 2002. Data mining for simple sequence repeats in expressed sequence tags from barley, maize, rice, sorghum and wheat. Plant Molecular Biology, 48 (5-6): 501–510.

Kirst, M.; Basten, C.J.; Myburg, A.A.; *et al.* 2005. Genetic architecture of transcript-level variation in differentiating xylem of a eucalyptus hybrid. Genetics, 169 (4): 2295–2303.

Kliebenstein, D.J. 2012. Exploring the shallow end; estimating information content in transcriptomics studies. Frontiers in Plant Science, 3: 213.

Koenig, D.; Jiménez-Gómez, J.M.; Kimura, S.; *et al.* 2013. Comparative transcriptomics reveals patterns of selection in domesticated and wild tomato. Proceedings of the National Academy of Sciences U.S.A., 110 (28): E2655–E2662.

Kumpatla, S.P.; Buyyarapu, R.; Abdurakhmonov, I.Y.; Mammadov, J.A. 2012. Genomics-assisted plant breeding in the 21st century: technological advances and progress. In: Abdurakhmonov, I. (ed.). Plant Breeding. Rijeka, Croatia: In Tech, pp. 131–184.

Libault, M.; Farmer, A.; Joshi, T.; *et al.* G. 2010. An integrated transcriptome atlas of the crop model Glycine max, and its use in comparative analyses in plants. The Plant Journal, 63 (1): 86–99.

Lindlöf, A. 2003. Gene identification through large-scale EST sequence processing. Applied Bioinformatics, 2 (3): 123–129.

Liu, C.; Li, S.; Wang, M.; Xia, G. 2012. A transcriptomic analysis reveals the nature of salinity tolerance of a wheat introgression line. Plant Molecular Biology, 78 (1-2): 159–169.

Luk'ianov, S.A.; Gurskaia, N.G.; Luk'ianov, K.A.; *et al.* 1994. Highly-effective subtractive hybridization of cDNA. Bioorganicheskaya khimiya, 20: 701–704.

Lukyanov, S.A.; Rebrikov, D.; Buzdin, A.A. 2007. Suppression subtractive hybridization. In: Buzdin, A.A. and Lukyanov S.A. (eds). Nucleic Acids Hybridization Modern Applications. Dordrecht, The Netherlands: Springer, pp. 53–84.

Madden, S. L.; Wang, C.J.; Landes, G. 2000. Serial analysis of gene expression: from gene discovery to target identification. Drug Discovery Today, 5 (9): 415–425.

Mafra, V.; Martins, P.K.; Francisco, C.S.; *et al.* 2013. *Candidatus Liberibacter americanus* induces significant reprogramming of the transcriptome of the susceptible citrus genotype. BMC Genomics, 14 (1): 247.

Martin, J.; Bruno, V.M.; Fang, Z.; *et al.* 2010. Rnnotator: an automated de novo transcriptome assembly pipeline from stranded RNA-Seq reads. BMC Genomics, 11 (1): 663.

Matsumura, H.; Nirasawa, S.; Terauchi, R. 1999. Technical advance: transcript profiling in rice (*Oryza sativa* L.) seedlings using serial analysis of gene expression. Plant Journal, 20: 719–726.

Matsumura, H.; Reich, S.; Ito, A.; *et al.* 2003. Gene expression analysis of plant host–pathogen interactions by SuperSAGE, Proceedings of the National Academy of Sciences U.S.A., 100: 15718–15723.

Mockler, T.C.; Chan, S.; Sundaresan, A.; *et al.* 2005. Applications of DNA tiling arrays for whole-genome analysis. Genomics, 85: 1–15.

Morozova, O.; Hirst, M.; Marra, M.A. 2009. Applications of new sequencing technologies for transcriptome analysis. Annual Review of Genomics and Human Genetics, 10: 135–151.

Nacht, M.; Ferguson, A.T.; Zhang, W.; et al. 1999. Combining serial analysis of gene expression and array technologies to identify genes differentially expressed in breast cancer. Cancer Research, 59: 5464–5470.

Nagalakshmi, U.; Wang, Z.; Waern, K.; et al. 2008. The transcriptional landscape of the yeast genome defined by RNA sequencing. Science, 320: 1344–1349.

Nagaraj, S.H.; Gasser, R.B.; Ranganathan S. 2006. A hitchhiker's guide to expressed sequence tag (EST) analysis. Briefings in Bioinformatics, 8 (1): 6–21.

Nielsen, K.L.; Hogh, A.L.; Emmersen, J. 2006. DeepSAGE–digital transcriptomics with high sensitivity, simple experimental protocol and multiplexing of samples. Nucleic Acids Research, 34: e133.

Novaes, E.; Drost, D.R.; Farmerie, W.G.; et al. 2008. High-throughput gene and SNP discovery in *Eucalyptus grandis*, an uncharacterized genome. BMC Genomics, 9 (1): 312.

Oshlack, A.; Robinson, M.D.; Young M.D. 2010. From RNA-seq reads to differential expression results. Genome Biology, 11: 220.

Palmieri, D.A.; Novelli, V.M.; Bastianel, M.; et al. 2007. Frequency and distribution of microsatellites from ESTs of citrus. Genetics and Molecular Biology, 30 (3): 1009–1018.

Pérez-de-Castro, A.M.; Vilanova, S.; Cañizares, J.; et al. 2012. Application of genomic tools in plant breeding. Current Genomics, 13: 179–195.

Rebrikov, D.V.; Britanova, O.V.; Gurskaya, N.G.; et al. 2000. Mirror orientation selection (MOS): a method for eliminating false positive clones from libraries generated by suppression subtractive hybridization. Nucleic Acids Research, 28: e90.

Redman, J.C.; Haas, B.J.; Tanimoto, G.; Town, C.D. 2004. Development and evaluation of an *Arabidopsis* whole genome Affymetrix probe array. The Plant Journal, 38: 545–561.

Robertson, G.; Schein, J.; Chiu, R.; et al.. 2010. De novo assembly and analysis of RNA-seq data. Nature Methods, 7 (11): 909–912.

Robinson, M.D.; McCarthy, D.J.; Smyth, G.K. 2010. EdgeR: a bioconductor package for differential expression analysis of digital gene expression data. Bioinformatics, 26: 139–140.

Rodrigues, C.M.; de Souza, A.A.; Takita, M.A.; et al. 2013. RNA Seq analysis of *Citrus reticulata* in the early stages of *Xylella fastidiosa* infection reveals auxin-related genes as a defense response. BMC Genomics, 14 (1): 676.

Rudd, S. 2003. Expressed sequence tags: alternative or complement to whole genome sequences? Trends in Plant Science, 8 (7): 321–329.

Saha, S.; Sparks, A.B.; Rago, C.; et al. 2002. Using the transcriptome to annotate the genome. Nature Biotechnology, 20: 508–512.

Salazar, M.M.; Nascimento, L.C.; Camargo, E.L.O.; et al. 2013. Xylem transcription profiles indicate potential metabolic responses for economically relevant characteristics of *Eucalyptus* species. BMC Genomics, 14 (1): 201.

Sanger, F.; Nicklen, S.; Coulson, A.R. 1977. DNA sequencing with chain terminating inhibitors. Proceedings of the National Academy of Sciences U.S.A., 74: 5463–5467.

Sato, Y.; Baltazar, AA.; Nobukazu, N.; *et al.* 2011. RiceXPro: a platform for monitoring gene expression in japonica rice grown under natural field conditions. Nucleic Acids Research, 39 (1): D1141–D1148.

Schena, M.; Shalon, D.; Davis, R.W.; Brown, P.O. 1995. Quantitative monitoring of gene expression patterns with a complementary DNA microarray. Science, 270 (5235): 467–470.

Schmid, M.; Davison, T.S.; Henz, S.R.; *et al.* 2005. A gene expression map of *Arabidopsis thaliana* development. Nature Genetics, 37 (5): 501–506.

Schuler, G.D.; Boguski, M.S.; Stewart, E.A.; *et al.* 1996. A gene map of the human genome. Science, 274: 540–546.

Shannon, P.; Markiel, A.; Ozier, O.; *et al.* 2003. Cytoscape: a software environment for integrated models of biomolecular interaction networks. Genome Research, 13 (11): 2498–2504.

Sharopova, N.; McMullen, M.D.; Schultz, L.; *et al.* 2002. Development and mapping of SSR markers for maize. Plant Molecular Biology, 48 (5-6): 463–481.

Shendure, J.; Ji, H. 2008. Next-generation DNA sequencing. Nature Biotechnology, 26 (10): 1135–1145.

Soneson, C.; Delorenzi, M. 2013. A comparison of methods for differential expression analysis of RNA-seq data. BMC Bioinformatics, 14 (1): 91.

Stoesser, G.; Baker, W.; van den Broek, A.; *et al.* 2003. The EMBL nucleotide sequence database: major new developments. Nucleic Acids Research, 31: 17–22.

Surget-Groba, Y.; Montoya-Burgos, J.I. 2010. Optimization of de novo transcriptome assembly from next-generation sequencing data. Genome Research, 20 (10): 1432–1440.

Swanson-Wagner, R.; Briskine, R.; Schaefer, R.; *et al.* 2012. Reshaping of the maize transcriptome by domestication. Proceedings of the National Academy of Sciences U.S.A., 109 (29): 11878–11883.

Thimm, O.; Bläsing, O.; Gibon, Y.; *et al.* 2004. Mapman: a user-driven tool to display genomics data sets onto diagrams of metabolic pathways and other biological processes. The Plant Journal, 37 (6): 914–939.

Tokimatsu, T.; Sakurai, N.; Suzuki, H.; *et al.* 2005. KaPPA-View. A web-based analysis tool for integration of transcript and metabolite data on plant metabolic pathway maps. Plant Physiology, 138 (3): 1289–1300.

Trapnell, C.; Pachter, L.; Salzberg, S.L. 2009. TopHat: discovering splice junctions with RNA-Seq. Bioinformatics, 25: 1105–1111.

Trapnell, C.; Williams, B.A.; Pertea, G.; *et al.* 2010. Transcript assembly and quantification by RNA-Seq reveals unannotated transcripts and isoform switching during cell differentiation. Nature Biotechnology, 28 (5): 511–515.

Trick, M.; Adamski, N.M.; Mugford, S.G.; *et al.* 2012. Combining SNP discovery from next-generation sequencing data with bulked segregant analysis (BSA) to fine-map genes in polyploid wheat. BMC Plant Biology, 12 (1): 14.

Trick, M.; Long, Y.; Meng, J.; Bancroft, I. 2009. Single nucleotide polymorphism (SNP) discovery in the polyploid *Brassica napus* using Solexa transcriptome sequencing. Plant Biotechnology Journal, 7: 334–346.

Varshney, R.K.; Glaszmann, J-C.; Leung, H.; Ribaut, J.-M. 2010. More genomic resources for less-studied crops. Trends in Biotechnology, 28: 452–460.

Velculescu, V.E.; Zhang, L.; Vogelstein, B.; Kinzler, K.W. 1995. Serial analysis of gene expression. Science-AAAS-Weekly Paper Edition, 270 (5235): 484–486.

Vega-Sánchez, M.E.; Malali, G.; Guo-Liang, W. 2007. Tag-based approaches for deep transcriptome analysis in plants. Plant Science, 173 (4): 371–380.

Venu, R.; Sreerekha, M.; Nobuta, K.; *et al.* 2011. Deep sequencing reveals the complex and coordinated transcriptional regulation of genes related to grain quality in rice cultivars. BMC Genomics, 12 (1): 190.

Vettore, A.L.; da Silva, F.R.; Kemper, E.L.; *et al.* 2003. Analysis and functional annotation of an expressed sequence tag collection for tropical crop sugarcane. Genome Research, 13 (12): 2725–2735.

Vidal, R.O.; Mondego, J.M.C.; Pot, D.; *et al.* 2010. A high-throughput data mining of SNPs in *Coffea* spp. suggests differential ESTs homeologous gene expression in the allotetraploid *Coffea arabica*. Plant Physiology (Bethesda), 154: 1053–1066.

Wang, T.; Chen, L.; Zhao, M.; *et al.* 2011. Identification of drought-responsive microRNAs in *Medicago truncatula* by genome-wide high-throughput sequencing. BMC Genomics, 12 (1): 367.

Wang, Z.; Gerstein, M.; Snyder, M. 2009. RNA-Seq: a revolutionary tool for transcriptomics. Nature Reviews Genetics, 10 (1): 57–63.

Wei, C.L.; Ng, P.; Chiu, K.P.; *et al.* 2004. 5′ Long serial analysis of gene expression (LongSAGE) and 3′ LongSAGE for transcriptome characterization and genome annotation. Proceedings of the National Academy of Sciences U.S.A., 101: 11701–11706.

Wong, M.L.; Medrano, J.F. 2005. Real-time PCR for mRNA quantitation. Biotechniques, 39 (1): 75.

Yamamoto, N.; Tsugane, T.; Watanabe, M.; *et al.* 2005. Expressed sequence tags from the laboratory-grown miniature tomato (*Lycopersicon esculentum*) cultivar Micro-Tom and mining for single nucleotide polymorphisms and insertions/deletions in tomato cultivars. Gene, 356: 127–134.

Yao, Y.; Ni, Z.; Zhang, Y.; *et al.* 2005. Identification of differentially expressed genes in leaf and root between wheat hybrid and its parental inbreds using PCR-based cDNA subtraction. Plant Molecular Biology, 58 (3): 367–384.

Zerbino, D.R.; Birney, E. 2008. Velvet: algorithms for de novo short read assembly using de Bruijn graphs. Genome Research, 18 (5): 821–829.

Zhang, L.; Zhou, W.; Velculescu, V.E.; *et al.* 1997. Gene expression profiles in normal and cancer cells. Science, 276: 1268–1272.

Zimmermann, P.; Hirsch-Hoffmann, M.; Hennig, L.; Gruissem, W. 2004. GENEVESTIGATOR. Arabidopsis microarray database and analysis toolbox. Plant Physiology, 136 (1): 2621–2632.

4 Proteomics

Ilara Gabriela F. Budzinski, Thaís Regiani, Mônica
T. Veneziano Labate, Simone Guidetti-Gonzalez,
Danielle Izilda R. da Silva, Maria Juliana Calderan
Rodrigues, Janaina de Santana Borges, Ivan Miletovic
Mozol, and Carlos Alberto Labate
*Department of Genetics, University of São Paulo/ESALQ, Piracicaba,
SP, Brazil*

History

The origins of plant breeding programs can be traced back to around
11 000 years ago with the domestication of wild plants. Plants with desirable characteristics were empirically selected to identify those that were
best adapted for agriculture (Bruins, 2010). Nowadays, plant breeding is
accomplished with the use of a broad spectrum of different techniques,
from phenotype selection to complex molecular methods, to aggregate
desirable characteristics to cultivars (Denis and Bouvet, 2013). In these
areas, proteomics can be a powerful tool for researchers who work with
plant breeding programs, because it is able to provide information, at the
molecular level, about the genetic variability actually expressed by the
genome. (Pennington and Dunn, 2001).

Proteomics emerged as one of the offshoots of the post-genomic era
and has since contributed significantly to a global, integrated understanding of biological systems. The study of the proteome involves the
entire set of proteins expressed by the genome of a cell, but it can also
be directed to cover just those proteins that are differentially expressed
under specific conditions (Meireles, 2007).

Initially, proteomics was associated with the identification of a large
number of proteins in a cell or organism, using isoelectric focusing
and two-dimensional gels (Meireles, 2007). The first proteomics-related
studies occurred in this context in the 1970s, when researchers began
to map the proteins in a few organisms and created the first data
bases for proteins (O'Farrel, 1975; Klose, 1975; Scheele, 1975). The term
"proteome", however, only appeared in 1994 to designate the group
of proteins expressed by the genome of a cell in a given environment
(Wilkins *et al.*, 1995; Anderson and Anderson, 1996).

Omics in Plant Breeding, First Edition. Edited by Aluízio Borém and Roberto Fritsche-Neto.
© 2014 John Wiley & Sons, Inc. Published 2014 by John Wiley & Sons, Inc.

Proteomics has been evolving mainly on the basis of protein separation by two-dimensional gel electrophoresis and chromatographic techniques (Eberlin, 2005). However, it was in mass spectrometry that proteomics found one of its principal allies. This technique is commonly used to analyze molecules in relation to their mass/charge (m/z) ratio. The procedure consists of ionizing the molecules to enable them to receive or donate protons and become positively or negatively charged; when they are subsequently exposed to an electric field, the ions will shift inside the analyzer according to the chemical characteristics of the molecules (see Chapter 8). At present, proteomics involves the functional analysis of gene products, a line of investigation conducted in a general area known as "functional genomics," which includes the large-scale identification, location, and compartmentalization of proteins, as well as the study and construction of protein interaction networks (Aebersold and Mann, 2003). The object of proteomics research is to help to attain a holistic view of a living organism, to understand its responses to stimuli, and ultimately to be able to predict biological events that may affect it.

The proteomic study of plant species has a variety of objectives, such as to assist plant breeding programs designed, for example, to identify proteins that cause allergies in humans (Schenk *et al.*, 2011), or to understand how the proteins are involved in the protection and response of plants subjected to biotic and abiotic stress (Kosová *et al.*, 2011).

Considering the exceptional importance of proteomics as an auxiliary tool for plant breeding programs, the following sections are designed to convey general information on total protein extraction methods, subproteomics, post-translational modifications, and future perspectives.

Different Methods for the Extraction of Total Proteins

Importance of High-quality Protein Extracts

The two-dimensional electrophoresis technique (2DE) is widely used in proteomic studies, particularly for comparative analyses between samples (Wang *et al.*, 2008). Although there are more advanced proteomic techniques that do not require the use of gel, 2DE is still the most readily available and commonly used for the separation of proteins (Görg and Weiss, 2004). For best quality results, the protein must be separated with high resolution, which requires great care in the preparation of the samples. Many problems during the isoelectric focusing and 2DE separation are caused by the presence of non-proteic components in the extract, which are capable of hindering the migration (Görg and Weiss, 2000). Because these organisms are plants, the protein extraction is the most challenging one because their tissues are rich in proteases and interferent components, such as phenolic compounds (Wang *et al.*, 2008).

The next sections describe some of the methods used to extract proteins from plants. Each of them has advantages and disadvantages; the method should be carefully chosen in accordance with the material that is being investigated, to ensure that the resulting extract is of the highest possible quality. Figure 4.1 shows the different methods that are commonly in use for the extraction of plant proteins.

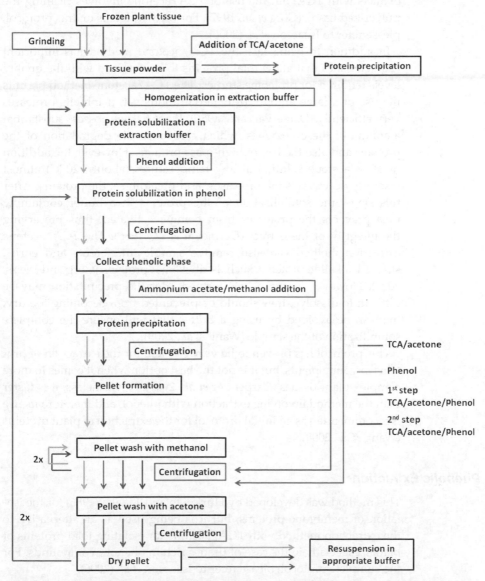

Figure 4.1   Diagram illustrating the three predominant methods for the extraction of plant proteins. (See color figure in color plate section).

Extraction with TCA/Acetone

This method was first developed by Damerval *et al.* (1986) to extract proteins from wheat seedlings, but it is now used extensively to precipitate proteins from tissues or cells by treatment with trichloroacetic acid (TCA, 10% m/v) in icy acetone (−20 °C). This extraction technique is based on protein denaturation in an acid and/or hydrophobic medium, which in turn helps the precipitation of proteins and the removal of interferent substances. Some protocols use direct precipitation of the protein extracts with TCA, but the use of TCA/acetone involves cleaning the pulverized tissue (Görg *et al.*, 1997). For further details on this protocol, please refer to Isaacson *et al.* (2006).

In addition to being fast, the TCA/acetone method has important characteristics that allow total proteins to be obtained with the quality levels required for proteome studies. The TCA/acetone method permits the use of a large amount of material; moreover, it inhibits proteases very efficiently (Damerval *et al.*, 1986). Plant tissues possess substantial amounts of these enzymes, which can cause the degradation of the proteins and also the loss of high-mass proteins. However, the addition of TCA/acetone solution at low temperatures (about −20 °C) almost instantly deactivates the proteases, and precipitates the proteins. After this stage, the solubilization of the proteins in a buffer containing urea prevents the proteases from being reactivated, thus preserving the integrity of the extract (Dermeval *et al.*, 1986). The TCA/acetone extraction method can also remove sample interferents and enrich strongly alkaline proteins, such as ribosomal proteins (Görg and Weiss, 2004). However, as the pellets formed after the precipitation may be difficult to dissolve, they should be prevented from becoming too dry; this can be avoided by using a buffer that will ensure the complete solubilization of the sample (Wang *et al.*, 2008).

This protocol has proven to be very efficient for the extraction of proteins in young plants, but it is not the best option when it comes to more complex plant tissues (Carpentier *et al.*, 2005). Nevertheless, it is faster than the method involving extraction with phenol, and therefore its use is recommended as an initial protocol for the extraction of plant proteins (Wang *et al.*, 2008).

Phenolic Extraction

This method was developed by Hurkman and Tanaka (1986) for the isolation of membrane proteins, but it is being used as an alternative to the extraction method with TCA/acetone for isolating total proteins of plants, especially in the case of tissues rich in phenolic compounds. For the complete protocol, please refer to Isaacson *et al.* (2006).

In this method, the ground material is mixed with the extraction buffer containing phenol, and the phenol fraction is then precipitated with methanol or acetone. The literature shows that the method can vary. In some cases, sucrose is added to the extraction buffer (Wang et al., 2003; Wang et al., 2006), which makes the aqueous phase denser than the phenol phase. As a result, when the mixture is centrifuged, the phenol phase separates and it can be collected in the upper part of the tube, becoming easier to be removed. The phenol dissolves proteins and lipids, while other substances remain dissolved in the lower phase. After the phases are separated, the proteins are simultaneously purified and concentrated by precipitation with methanol or acetone (Isaacson et al., 2006; Wang et al., 2003). The pellet that forms after the precipitation is usually white; a yellow coloration may indicate that phenolic compounds have precipitated along with the proteins. When this occurs, maceration of the material with polyvinylpolypyrrolidone (PVPP) is recommended to help to remove the phenolic compounds (Wang et al., 2003).

Despite the high efficiency of the phenolic extraction, care should be taken with a few details if high-quality extracts are to be produced successfully. During the first extraction phase, all the material should be kept at a low temperature (4 °C); the recovery of the phenolic phase also requires a great deal of care (Faurobert et al., 2007) to avoid contamination. Furthermore, the phenol to be used should be equilibrated at a pH value of 8 by adjustment with TRIS-HCl (Wang et al., 2008). As with TCA/acetone, the use of phenol helps to reduce protein degradation induced by endogenous proteolytic activity (Schuster and Davies, 1983). Also, many studies have shown that the use of the phenolic extraction method for recalcitrant tissues can efficiently eliminate compounds that may interfere in the separation by electrophoresis (Vincent et al., 2006).

The method of extraction with phenol is more efficient than the TCA/acetone method (Wang et al., 2008): the phenolic method is capable of producing highly pure samples, because contaminants such as polysaccharides and other water-soluble compounds are separated into different phases by centrifugation (Isaacson et al., 2006). It does, however, involve more work and take more time; in addition, phenol is very toxic and, therefore, must be handled with care. All these advantages and restrictions should be taken into careful consideration when selecting the methodology to be used in the investigation.

Extraction with Phenol and Precipitation with TCA/Acetone

This extraction method has the additional advantage of combining precipitation with TCA/acetone (to remove phenolic compounds, lipids, and pigments) and extraction with phenol (to remove phenolic compounds, lipids, pigments, polysaccharides, nucleic acids, and salts) (Wang et al., 2008). The use of previously TCA/acetone-precipitated tissue as a starting material for phenolic extraction was proposed by Wang et al. (2003) as an alternative method for obtaining a protein extract from leaves of olive trees (*Oleaeuropea* L.). This initial precipitation helps to remove interferents, which makes it easier to extract the proteins and prepare the 2DE gels. However, the possibility that the repeated washings may cause loss of protein should be taken into consideration.

Protein Extraction: Removal of Ribulose-1,5-Bisphosphate Carboxylase/Oxygenase (RuBisCO)

Most proteins detected by 2DE are of the very abundant type. This makes less abundant proteins, such as regulators and signaling factors, especially difficult to identify, mainly due to the presence of predominant proteins. The very abundant proteins limit not only the resolution and yield of the remaining proteins, but they can also mask and negatively affect the electrophoretic migration of neighboring proteins (Cho *et al.*, 2008).

In plant species, RuBisCO is the most abundant protein in green tissues (Widjaja *et al.*, 2009) and represents about 5.0–50% of the total proteins found in plants with C_4 and C_3 metabolism, respectively (Feller *et al.*, 2008). In this context, different methods are being developed for the reduction or complete removal of RuBisCO from leaf protein extracts. These protocols propose fractionation of the protein extract with calcium chloride ($CaCl_2$) and phytate (phytic acid) (Krishnan and Natarajan, 2009), extraction with Mg/Nonidet P-40 (NP-40) buffer, followed by fractionation with polyethylene glycol (PEG) (Acquadro *et al.*, 2009), or alterntively the use of specific columns for the removal of RuBisCO (Cellar *et al.*, 2008).

Subcellular Proteomics

The characterization of subcellular proteomes has been stimulating increasing interest and achieving substantial success. The object of subcellular fractionation prior to proteomic analysis is to separate the proteins present in a given organelle, in order to minimize contamination by other compartments (Feiz *et al.*, 2006). The importance of this technique was clearly demonstrated by the work of Marmagne *et al.* (2004), which showed that 95% of the membrane proteins identified by this method in *Arabidopsis* had not been described before.

Several methods are being used for the analysis of membrane proteins. Most of them consist of maceration of the plant tissue in a buffer containing salts, followed by filtration and subsequent separation of the fractions by centrifugation. An evaluation of plasma membrane proteins is expected to provide evidence of how the primary structure of the cell operates with regard to ion and metabolite transport, endocytosis and differentiation, and cell proliferation. All of these operations involve a series of proteins with highly diversified functions and structures (Komatsu *et al.*, 2007). Cytoplasm membrane transporters, for example, are related to a tolerance to excess of water; therefore, studies of these proteins may lend support to improvement work conducted with the purpose of obtaining plants that would be resistant to this stress condition (Shabala, 2011). Proteins involved in metabolism, energy, and defense have been found in membranes of the Golgi complex in rice (Tanaka *et al.*, 2004). In their investigations, also with rice, Komatsu *et al.* (2007) observed changes in vacuolar proteins that are involved in mechanisms related to root growth (Komatsu *et al.*, 2007).

In nuclear proteomic studies, proteins are separated by fractionation based on physical differences between organelles. The first step consists of a filtration followed by short, low-speed centrifugation with or without the aid of density gradients (Percoll or sucrose) (Agrawal *et al.*, 2011). Additional centrifugations are used to increase the purity of the enriched organelles (Silva, 2012). The purity of nuclear protein extracts may be checked by means of the Western blot technique (Agrawal *et al.*, 2011). Nuclear protein databases, such as NMPdb (Mika and Rost, 2005), predictNLS and NucPred (Fink *et al.*, 2008) are already available, and their improvement will enhance the development of subcellular proteomics (Erhardt *et al.*, 2010). The nuclear proteome is dynamic and responds to environmental and intracellular stimuli (Pandey *et al.*, 2008). The nucleus is the most prominent structure in eukaryotic cells, and a good knowledge of its molecular configuration will help to accomplish one of the most important goals of cell biology (Erhardt *et al.*, 2010). The cell nucleus contains proteins related to regulatory and signaling factors (Repetto *et al.*, 2008), as well as proteins that act on the metabolism of RNAs, on the organization of cell structures (Blazek and Meisterernst, 2006), and on the organization of the DNA for gene replication and expression (Abdalla and Rafudeen, 2012). Therefore, understanding and learning to manipulate such proteins is widely applicable in plant improvement. Studies conducted with rice (Choudhary *et al.*, 2009), sugarcane (Silva, 2012), and chickpea (Pandey *et al.*, 2008) stand out in this field of research.

Mitochondria are also isolated by fractionation. After the lysis, the cell fragments are centrifuged at low speed for the separation of other organelles. The supernatant is then centrifuged at high speed, and the resulting fraction is purified by means of additional gradient centrifugation. The purity of the fraction may be assessed with the use of a number of dyes, such as Rhodamine 123 (Agrawal *et al.*, 2011). The research efforts in this area relate to the study of the mitochondria proteome to determine its composition under different conditions, such as programmed cell death under conditions of saline stress (Chen *et al.*, 2009).

Chloroplasts are found in plants and some single-cell eukaryotes that are able to conduct photosynthesis (Dreger, 2003), so the study of their proteome becomes fundamental to understanding this process. The isolation of this organelle also follows the methodology of fractionation and centrifugation with or without density gradients. The chloroplastidial proteome has been studied under different environmental conditions, such as exposure to salinity (Zörb *et al.*, 2009) and tolerance to dryness (Kosmala *et al.*, 2012), as well as with regard to its biosynthetic routes, such as the isoprenoid pathways (Joyard *et al.*, 2009), among others.

The extraction of cell wall proteins has been accomplished with rice (Chen *et al.*, 2009), corn (Zhu *et al.*, 2006), *Arabidopsis* (Feiz *et al.*, 2006), and sugarcane (Calderan-Rodrigues, 2014). There are two major types of extraction of cell wall proteins: disruptive extraction and non-disruptive extraction. The former ruptures the cell wall to promote the contact of the buffers with the proteins; while this makes it possible to isolate and remove the proteins, it may also lead to increased contamination by proteins other than wall proteins. The non-disruptive extraction, on the other hand, is designed to keep the cells alive, but the buffers have only limited access to the proteins. In *Arabidopsis*, about 400 wall proteins have been identified, such as those involved in carbohydrate and oxydo-reductase metabolism (Jamet *et al.*, 2008). Targeting a commercial application, Lionetti *et al.* (2010) succeeded in altering pectins in *Arabidopsis*, tobacco, and wheat by means of the expression of a number of proteins. This experiment increased the efficiency of enzymatic saccharification, a technique which is used in the production of ethanol from cellulose – and, as such, is one of the main obstacles for reducing the costs of this technology to make it more viable.

In addition to identifying the expected proteins in a cell compartment, subcellular proteome studies can prompt the discovery of new proteins or the discovery of unknown functions in already known proteins, and thus broaden the way to genetic improvement. Therefore, organelle proteomics not only allows the understanding of the processes that take place in a given compartment, but it also leads to an overall understanding of the mechanisms of plant growth and development.

Post-Translational Modifications

Post-translational modifications (PTMs) of proteins are covalent molecular changes that define the activity, location, turnover, and interaction of a protein with other proteins (Mann and Jensen, 2003). In particular, cysteine, asparagine, serine, threonine, tyrosine, and lysine side chains may be modified by the addition of a large variety of molecules – mostly carbohydrates, lipids, and phosphate, methyl or nitrate groups (Bendixen, 2013). Thereby, PTMs give the proteome an enormous diversity, allowing the cell to respond flexibly to different stimuli (Howden and Huitema, 2012).

PTMs include phosphorylation, glycosylation, ubiquitination, methylation, and acetylation, among others. More than 300 modification types have already been identified and may be classed into four groups:

(a) addition of functional groups, (b) addition of proteins or peptides, (c) structural changes in proteins, and (d) changes in the chemical nature of an amino acid (Ytterberg and Jensen, 2010). These changes depend on molecular signals and may be affected by the development stage, location and/or biotic or abiotic factors, such as stress conditions and disease (Webster and Thomas, 2012). PTMs may change the chemical state of a protein statically or dynamically, and in such a subtle manner that it may not be easily detected by standard protein profiling techniques that use gel (Simon and Cravatt, 2008).

Recent advances in the field of PTMs have revealed their physiological relevance and universal occurrence. This makes the elucidation of these modifications one of the most challenging themes in proteomics (Schulze, 2010), because PTMs often occur in stoichiometric levels that are too low for comprehensive analyses. Consequently, different approaches are required to enrich peptides with PTMs, as well as to develop specific analytical procedures, without which such modifications may not be properly sampled and might therefore be neglected (Lange and Overall, 2013).

An understanding of the dynamics of the expression of protein PTMs, combined with the study of their functions, may help to obtain stress-tolerant cultures endowed with new characteristics, using selection by biomarkers and/or genetic modification. Since phosphorylation is one of the most important PTMs in plants, it will now be discussed in detail.

Phosphorylation

Protein phosphorylation is one of the main PTMs in plants, especially those PTMs affecting cell regulation. Phosphorylations are catalyzed by protein kinases (PKs) and can cause changes in protein functions, such as enzymatic activity, subcellular location, and capacity to interact with ligands or other proteins. In certain cases, the phosphorylation process may also affect the stability of the protein (Nakagami *et al.*, 2012).

Phosphorylation of proteins increases the molecular mass of phosphotyrosine (pT) by 80 Da with addition of HPO_3; the locus of amino acid phosphorylation may be identified by a gain in mass of 80 Da (or 98 Da for H_3PO_4) in peptide ion fragments. Peptides containing phosphorylated serine and treonine undergo cleavage, through which they lose a phosphoric acid molecule (H_3PO_4) during the fragmentation process by (MS/MS) (Angel *et al.*, 2012).

Using a 2DE gel, the phosphoproteins can be detected with the aid of fluorescent dyes, such as ProQ-Diamond (Pro-Q DPS, Molecular Probes) and PhosTag™ (PerkinElmer), which bind selectively to the phosphate radical in the proteins (Pechanova *et al.*, 2013). Although the approaches based on 2DE gel are applied to phosphoproteomic characterization, the coverage by this technique is very limited.

The recent appearance of the "shotgun" proteomics dramatically amplified the proteome and the phosphoproteome coverage. As a result, thousands of proteins can be identified simultaneously from a complex sample, and the analyses have become much more practical (Nakagami *et al.*, 2012).

As the digestion of protein mixtures produces large numbers of differentially abundant products, phosphopeptides are impossible to detect without previously reducing the complexity of the sample. In other words, the phosphoproteins and/or phosphopeptides first need to be selectively concentrated (Nakagami *et al.*, 2012). Phosphopeptide enrichment using immobilized metal ion (Fe^{3+} or Ga^{3+}) affinity chromatography (IMAC) and metal oxide affinity chromatography (MOAC) with TiO_2 (Figure 4.2) or ZrO_2 is used extensively in phosphoproteomic studies (Aryal and Ross, 2010).

In recent years, many plant protein phosphorylation sites have been identified mostly through large-scale experiments using mass spectrometry (Chen *et al.*, 2010; Engelsberger and Schulze, 2012; Mayank *et al.*, 2012) and different stress conditions (Kline *et al.*, 2010; Engelsberger and Schulze, 2012; Shen *et al.*, 2012).

The importance of phosphoproteins in response to different stress conditions is being studied extensively in plants, particularly with regard to stress caused by water deficit (Bonhomme *et al.*, 2012), heat (Chen *et al.*, 2011), cold (Schulze *et al.*, 2012), salinity (Chang *et* al., 2012), and attack by pathogens (Benschop *et al.*, 2007; Serna-Sanz *et al.*, 2011).

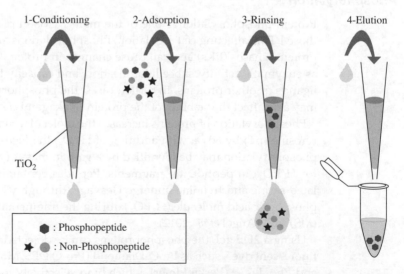

Figure 4.2 Enrichment of phosphopeptides through TiO_2 columns, showing the different steps of the procedure. Modified from "Using GL Sciences' Titansphere® Phos-TiO Kit" (www.glsciencesinc.com). (See color figure in color plate section).

In maize, 138 phosphopeptides have been identified and reported to show significant changes in expression, in response to the hydration regimen to which the plants were subjected. This made it possible to identify specific response patterns, which vary with the changes in the water status of the plant (Bonhomme *et al.*, 2012). Another study has shown that, following the perception of microbial signals, kinases and phosphatases act on specific proteins to modify complex signaling cascades. As a consequence, the system triggers a series of rapid defense responses (Jayaraman *et al.*, 2012).

Advances are also being made in quantitative analyses designed to map phosphorylation sites. Several strategies regarding quantitative proteomics have also been developed to obtain information on the dynamics of phosphoproteins; such strategies include label-free and chemical tagging quantification, or labeling with stable isotopes (Schulze and Usadel, 2010).

Recent breakthroughs in mass spectrometry technology have also been increasing the capacity for identifying protein phosphorylation sites on a large scale, phosphoproteomics being one of the areas of outstanding progress. The characterization of plant phosphoproteomes will provide an insight into their major regulatory systems, and thus help to improve agronomically important plants.

Quantitative Proteomics

Proteomics based on mass spectrometry (Aebersold and Mann, 2003) has been contributing enormously to the understanding of fundamental biological processes, which has led to the development of methods that are capable of quantifying thousands of proteins. The quantification of the differences between two or more physiological stages in a given biological system is very important, but it involves the use of challenging techniques. Even with the still widespread use of the classical 2DE and blot staining differentiation methods aided by DIGE-fluorescence (Ünlü *et al.*, 1997), mass spectrometry based on quantitative methods have grown considerably over the past few years.

The next section describes the most common techniques used in quantitative proteomics and draws attention to the importance of recently developed methods and their applications.

Quantification with Isotope Labeling of Proteins and Peptides

Peptide and protein labeling in quantitative proteomics may be metabolic or chemical. Metabolic labeling is based on the tagging of peptides with stable isotopes. This can be used for the quantification of proteins, since the physicochemical properties of labeled and natural proteins, as well as their mass spectrum signals, are very similar, if not identical (Ong *et al.*, 2002). Consequently, the absolute and relative quantifications of a sample of interest are made by comparison.

Introduced in about 2000, SILAC, a technique used for labeling amino acids with stable isotopes in cell culture (Ong *et al.*, 2002), now ranks among the best known methods, particularly for its application to animal cells. Its underlying principle lies in the addition of isotopically labeled arginine and lysine (^{13}C, ^{15}N) to the culture medium of the cells, so that most of the peptides resulting from the tryptic digestion contain at least one labeled amino acid (Ong *et al.*, 2002). The relative quantification is obtained by means of a comparative analysis of the intensity of the mass spectrum produced by the labeled and the unlabeled peptides (Figure 4.3). Variations of the technique include pulsed-SILAC (Wu *et al.*, 2012) and super-SILAC (Geiger *et al.*, 2011).

The use of ^{15}N is not very frequent in proteomics. Although it is used for the quantification of proteins in microorganisms such as bacteria (Soufi *et al.*, 2010) and yeasts (Ross *et al.*, 2004), it is scarcely employed

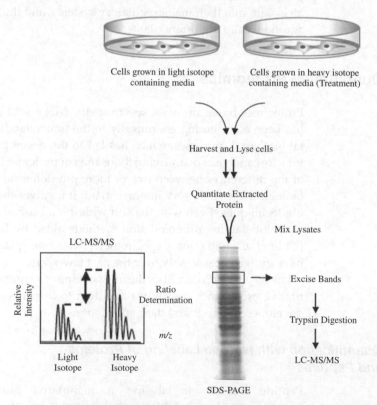

Figure 4.3 Schematic of SILAC Workflow, adapted from Thermo Scientific Pierce catalog for Mass Spectrometry Sample Preparation–V2, 2010. (See color figure in color plate section).

in plant proteomics and other higher, multicellular, organisms (Gouw et al., 2010). Since the number of labeled ^{15}N varies from one peptide to the next, its use makes the resulting data difficult to analyze, as opposed to the SILAC technique, which requires the addition of a fixed number of peptides – normally a single lysine or arginine.

In the case of chemical tagging of proteins and peptides, the technique is based on the principle that enzymatic reaction sites are modified by isotopic labeling. However, the targets of more recent techniques are lysine N-terminals and ε-amino groups of the peptides and proteins (Bantscheff et al., 2012). Among the existing chemical tagging methods, the best known are those that involve tandem mass tags (TMTs) and isobaric tags for relative as well as absolute quantification (iTRAQ). As both these methods target primary amines, they are called "isobaric tags" (Wiese et al., 2007). In this case, peptides from different tagged samples have identical mass (isobaric), but they can still be differentiated, after the mass spectrometry fragmentation, by the isotopic differential coding of the reporter ion, in a low-mass range, in-tandem spectrum MS (MS/MS), the intensity of which is used for the quantification. Another method of interest involves tagging the dimethyl group of the stable isotope, which represents an alternative for the direct chemical tagging of protein digestion products (Boersema et al., 2009).

Reverse-phase liquid chromatography systems directly coupled to mass spectrometers constitute the most sophisticated quantification technique currently in use. In this case, the resolution capability and the peak intensity result from the use of long columns (>20 cm) packed with particles under 3 µm, with elution gradients of about 2 h (Eeltink et al., 2009). This type of chromatography requires a high-pressure system (LC-UPLC) (Motoyama et al., 2006). Among the instruments that are now most commonly employed for quantitative proteomics and allowing characterization of proteomes of the order of 5000–10 000 proteins, Orbitraps, Q-TOFs, and triple quadrupoles may be cited as examples. In the majority of instances, this high-resolution and high-intensity capacity are associated with the use of an electrospray (ESI) ionization system as an interface for the chromatographic systems, and with the dissociation induced by high energy collision (HCD), a combination that works as a peptide fragmentation technique (Kim and Pandey, 2012) (for more details, please refer to Chapter 8). All these quantification methods are highly selective and capable of generating data with coefficients of variation lower than 20%, depending on the complexity of the sample under analysis (Bantscheff et al., 2012).

Label-free Quantification

The pioneering work of Washburn et al. (2001) for the large-scale, multidimensional separation of peptides for the identification of proteins showed the number of identified peptides to be correlated with a given protein, which suggests that the absolute quantification of the protein would be possible without the need to use reference standards labeled with stable isotopes. Since then, the technology for the identification and quantification of proteins without the use of amino acid labeling has been growing at a remarkable pace (Nahnsen *et al.*, 2013).

The techniques are based on two types of quantification methods: those that use the intensity of a full scan MS signal from the intact peptide for the direct quantification of the equivalent protein (Timm *et al.*, 2008), and those that infer the quantification indirectly by counting the spectra obtained from each protein (Zhang *et al.*, 2006).

The methods that use the intensity of the MS signal are more commonly employed for absolute quantifications (Schwanhausser *et al.*, 2011). High3 (or Top3) is an alternative quantification method (Silva *et al.*, 2006). In this case, the most intense signal from three specific peptides in a protein is taken into consideration, with the assumption that the intensity of the MS signal from such peptides is approximately the same. Even though both methods are efficient, spectral counting has been shown to be more dynamic and reproducible in the quantification; moreover, it allows the relative quantification of the abundance of proteins in a group of samples to be obtained (Washburn *et al.*, 2001). When working with large numbers of samples, however, the technique based on spectral counting requires a larger number of experimental parameters of the instrument to be taken into account, such as the exclusion of precursor ions and the width of the chromatographic peak, as well as specific parameters of the protein and peptides produced by proteolytic digestion.

As a rule, larger proteins generate larger numbers of peptides and, consequently, more spectral counting than smaller, but equally abundant, proteins. Therefore, the normalization of the spectral counting is very important for a precise quantification (Rappsilber *et al.*, 2002). The abundance and concentration of the samples should be given special attention to prevent them from becoming limiting factors in the dynamics of the detection of protein expression (Asara *et al.*, 2008). The limits of spectral counting and the level of confidence in their identity, relative to a given peptide, are particularly important if an analysis of quantitative proteomics shows statistically significant results with a probability of 95 or 99% (Cooper, 2010; Zhou *et al.*, 2010).

Perspectives

At present, the aspiration of those who work with proteomics is a ready availability of equipment associated with analytical techniques characterized by high sensitivity, high resolution, and high throughput levels. High sensitivity and high resolution are important, because most of the studies of interest involve proteins that are not particularly abundant in comparison with the total proteins in a complex biological sample. Given the general interest in studying the greatest possible number of proteins in a large number of biological samples, all at once, quickly, reproducibly and consistently, high-speed analysis and data processing are essential. As a result, the capacity of the analytical equipment has grown substantially over recent decades. The LC-MS instruments available today are much more sensitive than the equipment that was being marketed five years ago.

Equipment innovations are now developed in conjunction with bioinformatics and statistical analysis programs associated with the optimization of protocols for the extraction, digestion, and purification of protein samples. Another important step is the application of mass spectrometry to the construction of images of cells or tissue slides using the spatial distribution of the proteins or peptides.

Nevertheless, an ever more pervasive viewpoint is that the generation of proteomic data is, in itself, unable to resolve the complexity of biological systems. Moreover, concerns are being voiced about the organism as a whole, which raises the necessity of integrating the data obtained by means of the different molecular techniques used in genomics, transcriptomics, and metabolomics.

References

Abdalla, K.O.; Rafudeen, M.S. (2012). Analysis of the nuclear proteome of the resurrection plant *Xerophyta viscosa* in response to dehydration stress using iTRAQ with 2DLC and tandem mass spectrometry. Journal of Proteomics, 75 (8): 2361–2374.

Acquadro, A.; Falvo, S.; Mila, S.; *et al.* 2009. Proteomics in globe artichoke: Protein extraction and sample complexity reduction by PEG fractionation. Electrophoresis, 30: 1594–1602.

Aebersold, R.H.; Mann, M. 2003. Mass spectrometry-based proteomics. Nature, 422 (6928): 198–207.

Agrawal, G.K.; Bourguignon, J.; Rolland, N.; *et al.* 2011. Plant organelle proteomics: collaborating for optimal cell function. Mass Spectrometry Reviews, 30: 772–853.

Anderson, N.G.; Anderson, N.L. 1996. Twenty years of two-dimensional electrophoresis: past, present and future. Electrophoresis, 17 (3): 443–453.

Angel, T.E.; Aryal, U.K.; Hengel, S.M.; *et al.* 2012. Mass spectrometry-based proteomics: existing capabilities and future directions. Chemical Society Reviews, 41: 3912–3928.

Aryal, U.K.; Ross, A.R. 2010. Enrichment and analysis of phosphopeptides under different experimental conditions using titanium dioxide affinity chromatography and mass spectrometry. Rapid Communication in Mass Spectrometry, 24: 219–231.

Asara, J.M.; Christofk, H.R.; Freimark, L.M.; Cantley, L.C. 2008. A label-free quantification method by MS/MS TIC compared to SILAC and spectral counting in a proteomics screen. Proteomics, 8 (5): 994–999.

Bantscheff, M.; Schirle, M.; Sweetman, G.; *et al.* 2007. Quantitative mass spectrometry in proteomics: a critical review. Analytical and Bioanalytical Chemistry, 389 (4): 1017–1031.

Bantscheff, M.; Lemeer, S.; Savitski, M.M.; Kuster, B. 2012. Quantitative mass spectrometry in proteomics: critical review update from 2007 to the present. Analytical and Bioanalytical Chemistry, 404: 939–965.

Benschop, J.J.; Mohammed, S.; O'Flaherty, M.; *et al.* 2007. Quantitative phosphoproteomics of early elicitor signaling in *Arabidopsis*. Molecular and Cellular Proteomics, 6: 1198–1214.

Bendixen, E. 2003. Understanding the proteome. In: Toldrá F. and L M L Nollet, L.M.L. (eds). Proteomics in Foods-Principals and Applications, Food Microbiology and Food Safety, 2. New York: Springer Science + Business Media, chap. 1, pp. 3–19.

Blazek, E.; Meisterernst, M. 2006. A functional proteomics approach for the detection of nuclear proteins based on derepressed importin alpha. Proteomics, 6 (7): 2070–2078.

Boersema, P.J.; Raijmakers, R.; Lemeer, S.; *et al.* 2009. Multiplex peptide stable isotope dimethyl labeling for quantitative proteomics. Nature Protocols, 4(4): 484–494.

Bonhomme, L.; Valot, B.; Tardieu, F.; Zivy, M. 2012. Phosphoproteome dynamics upon changes in plant water status reveal early events associated with rapid growth adjustment in maize leaves. Molecular and Cellular Proteomics, 11: 957–972.

Bruins, M. 2010. A contribuição do melhoramento vegetal para a agricultura. Seednews, Jan/Feb 2010, XIV (1).

Calderan-Rodrigues, M.J.; Jamet, E.; Bonassi, M.B.C.R.; *et al.* 2014. Cell wall proteomics of sugarcane cell suspension cultures. Proteomics, 14: 738–749.

Carpentier, S.; Witters, E.; Laukens, K.; *et al.* 2005. Preparation of protein extracts from recalcitrant plant tissues: An evaluation of different methods for two dimensional gel electrophoresis analysis. Proteomics, 5: 2497–2507.

Cellar, N.A.; Kuppannan, K.; Langhorst, M.L.; *et al.* 2008. Cross species applicability of abundant protein depletion columns for ribulose-1,5- bisphosphate carboxylase/oxygenase. Journal of Chromatography, B, 861: 29–39.

Chang, I.F.; Hsu, J.L.; Hsu, P.H.; *et al.* 2012. Comparative phosphoproteomic analysis of microsomal fractions of *Arabidopsis thaliana* and *Oryza sativa* subjected to high salinity. Plant Science, 185–186: 131–142.

Chen, X.Y.; Kim, S.T.; Cho, W.K.; *et al.* 2009. Proteomics of weakly bound cell wall proteins in rice calli. Journal of Plant Physiology, 166 (7): 675–685.

Chen, X.; Wang, Y.; Li, J.; *et al.* 2009. Mitochondrial proteome during salt stress-induced programmed cell death in rice. Plant Physiology and Biochemistry, 47: 407–415.

Chen, X.; Zhang, W.; Zhang, B.; *et al.* 2011. Phosphoproteins regulated by heat stress in rice leaves. Proteome Science, 9 (37).

Chen, Y.; Höhenwarter, W.; Weckwerth, W. 2010. Comparative analysis of phytohormone – responsive phosphoproteins in *Arabidopsis thaliana* using TiO$_2$-phosphopeptide enrichment and MAPA. Plant Journal, 63: 1–17.

Cho, J.-H., Hwang, H., Cho, M.-H., *et al.* 2008. The effect of DTT in protein preparations for proteomic analysis: removal of a highly abundant plant enzyme, ribulose bisphosphate carboxylase/oxygenase. Journal of Plant Biology, 51: 297–301.

Choudhary, M.K.; Basu, D.; Datta, A.; *et al.* 2009. Dehydration-responsive nuclear proteome of rice (*Oryza sativa* L.) illustrates protein network, novel regulators of cellular adaptation, and evolutionary perspective. Molecular and Cellular Proteomics, 8: 1579–1598.

Cooper, B.; Feng, J.; Garrett, W. M. 2010. Relative, label-free protein quantitation: Spectral counting error statistics from nine replicate MudPIT samples. Journal of the American Society for Mass Spectrometry, 21 (9): 1534–1546.

Damerval, C.; De Vienne, D.; Zivy, M.; Thiellement, H. 1986. The technical improvements in two-dimensional electrophoresis increase the level of genetic variation detected in wheat-seedling proteins. Electrophoresis, 7: 52–54.

Denis. M.; Bouvet, J.M. 2013. Efficiency of genomic selection with models including dominance effect in the context of Eucalyptus breeding. Tree Genetics & Genomes, 9: 37–51.

Dreger, M. 2003. Subcellular proteomics. Mass Spectrometry Reviews, 22: 27–56.

Eberlin, M. A proteomics e os novos paradigmas. 2005. Jornal da Unicamp (University of Campinas, Brazil newspaper), 14–27 November 2005, p. 3.

Eeltink, S.; Dolman, S.; Swart, R.; *et al.* 2009. Optimizing the peak capacity per unit time in one-dimensional and off-line two-dimensional liquid chromatography for the separation of complex peptide samples. Journal of Chromatography A, 1216 (44): 7368–7374.

Engelsberger, W.R.; Schulze, W.X. 2012. Nitrate and ammonium lead to distinct global dynamic phosphorylation patterns when resupplied to nitrogen starved *Arabidopsis* seedlings. Plant Journal, 69: 978–995.

Erhardt, M.; Adamska, I.; Franco, O.L. 2010. Plant nuclear proteomics – inside the cell maestro. The FEBS Journal, 277: 3295–3307.

Faurobert, M.; Pelpoir, E.; Chaïb, J. 2007. Phenol extraction of proteins for proteomic studies of recalcitrant plant tissues. Methods in Molecular Biology, 355: 9–14.

Feiz, L.; Irshad, M.; Pont-Lezica, R.F.; *et al.* 2006. Evaluation of cell wall preparations for proteomics: a new procedure for purifying cell walls from *Arabidopsis* hypocotyls. Plant Methods, 2: 10–23.

Feller, U.; Anders, I.; Mae, T. 2008. Rubiscolytics: fate of Rubisco after its enzymatic function in a cell is terminated. Journal of Experimental Botany, 59: 1615–1624.

Fink, J.L.; Karunaratne, S.; Mittal, A.; *et al.* 2008. Towards defining the nuclear proteome. Genome Biology, 9: R15.

Geiger, T.; Wisniewski, J.R.; Cox, J.; *et al.* 2011. Use of stable isotope labeling by amino acids in cell culture as a spike-in standard in quantitative proteomics. Nature Protocols, 6 (2): 147–157.

Görg A.; Obermaier C.; Boguth G.; *et al.* 1997. Very alkaline immobilized pH gradients for two-dimensional electrophoresis of ribosomal and nuclear proteins. Electrophoresis, 18: 328–37.

Görg, A.; Weiss, W. 2000. 2D electrophoresis with immobilized pH gradients. In: Proteome Research: Two-Dimensional Electrophoresis and Identification Methods. Berlin, Heidelberg, New York: Springer, pp. 57–106.

Görg, A.; Weiss, W. 2004. Current two-dimensional electrophoresis technology for proteomics., Proteomics, 12: 3665–3685.

Gouw, J.W.; Krijgsveld, J.; Heck, A.J. 2010. Quantitative proteomics by metabolic labeling of model organisms. Molecular and Cellular Proteomics, 9 (1): 11–24.

Howden, A.J.M.; Huitema, E. 2012. Effector-triggered post-translational modifications and their role in suppression of plant immunity. Frontiers in Plant Science, 3: 160.

Hurkman, W.J.; Tanaka, C.K. 1986. Solubilization of plant membrane proteins for analysis by two-dimensional gel electrophoresis. Plant Physiology, 81: 802–806.

Isaacson, T., Saravanan, R.S., He, Y., *et al*. 2006. Sample extraction techniques for enhanced proteomic analysis of plant tissues. Nature Protocols, 1: 769–774.

Jamet, E.; Canut, H.; Albenne, C.; *et al*. 2008. Cell wall. In: Plant Proteomics: Technologies, Strategies, and Applications. Agrawal, G.K. and Rakwal, R. (eds). Hoboken: John Wiley & Sons, Inc., pp. 293–307.

Jayaraman, D.; Forshey, K.L.; Grimsrud, P.A.; Ané, J.M. 2012. Leveraging proteomics to understand plant–microbe interactions. Front Plant Science, 3: 44.

Joyard, J.; Ferro, M.; Masselon, C.; *et al*. 2009. Chloroplast proteomics and the compartmentation of plastidial isoprenoid biosynthetic pathways. Molecular Plant, 2 (6): 1154–1180.

Kim, M.S.; Pandey, A. 2012. Electron transfer dissociation mass spectrometry in proteomics. Proteomics, 12: 530–542.

Kline, K.G.; Barrett-Wilt, G.A.; Sussman, M.R. 2010. Plant changes in protein phosphorylation induced by the plant hormone abscisic acid. Proceedings of the National Academy of Sciences U.S.A., 107: 15986–15991.

Klose, J. 1975. Protein mapping by combined isoeletric focusing and electrophoresis of mouse tissues. A novel approach to testing for induced point mutations in mammals. Humangenetik, Berlin, 26 (3): 231–243.

Komatsu, S.; Konishi, H.; Hashimoto, M. 2007. The proteomics of plant cell membranes. Journal of Experimental Botany, 58 (1): 103–112.

Kosmala, A.; Perlikowski, D.; Pawłowicz, I.; Rapacz, M. 2012. Changes in the chloroplast proteome following water deficit and subsequent watering in a high- and a low-drought-tolerant genotype of *Festuca arundinacea*. Journal of Experimental Botany, 63 (17): 6161–6172.

Kosová, K.; Vítámvás, P.; Prášil, I. T.; Renaut, J. 2011. Plant proteome changes under abiotic stress – Contribution of proteomics studies to understanding plant stress response. Review Journal of Proteomics, 74: 1301–1322.

Krishnan, H.B.; Natarajan, S.S. 2009. A rapid method for depletion of Rubisco from soybean (Glycine max) leaf for proteomic analysis of lower abundance proteins. Phytochemistry, 70: 1958–1964.

Lange, P.F.; Overall, C.M. 2013. Protein TAILS: when termini tell tales of proteolysis and function. Current Opinion in Chemical Biology, 17: 73–82.

Lionetti, V.; Francocci, F.; Ferrari, S.; *et al.* 2010. Engineering the cell wall by reducing de-methyl-esterified homogalacturonan improves saccharification of plant tissues for bioconversion. PNAS, 107 (2): 616–621.

Mann, M.; Jensen, O.N. 2003. Proteomic analysis of post-translational modifications. Nature Biotechnology, 21: 255–261.

Marmagne, A.; Rouet, M.-A.; Ferro, M.; *et al.* 2004. Identification of new intrinsic proteins in *Arabidopsis* plasma membrane proteome. Molecular and Cellular Proteomics, 3 (7): 675–691.

Mayank, P.; Grossman, J.; Wuest, S.; *et al.* 2012. Characterization of the phosphoproteome of mature *Arabidopsis* pollen. Plant Journal, 72: 89–101.

Meireles, K.G.X. 2007. Aplicações da Proteomics na Pesquisa Vegetal. Embrapa. Documento 165, September, p. 41, http://www.infoteca.cnptia.embrapa.br/handle/doc/326868 (accessed 6 February 2013).

Mika, S.; Rost, B. 2005. NMPdb: Database of nuclear matrix proteins. Nucleic Acids Research, 33: D160–D163.

Motoyama, A.; Venable, J.D.; Ruse, C.I.; Yates, J.R. 2006. Automated ultra-high-pressure multidimensional protein identification technology (UHP-MudPIT) for improved peptide identification of proteomic samples. Analytical Chemistry, 78 (14): 5109–5118.

Nahnsen, S.; Bielow, C.; Reinert, K.; Kohlbacher, O. 2013. Tools for label-free peptide quantification. Molecular and Cell Proteomics, 12: 549–556.

Nakagami, H.; Sugiyama, N.; Ishihama, Y.; Shirasu, K. 2012. Shotguns in the front line: Phosphoproteomics in plants. Plant Cell Physiology, 53: 118–124.

O'Farrel, P.H. 1975. High resolution two-dimensional electrophoresis of proteins. The Journal of Biological Chemistry, 250 (10): 4007–4021.

Ong, S.E.; Blagoev, B.; Kratchmarova, I.; *et al.* 2002. Stable isotope labeling by amino acids in cell culture, SILAC, as a simple and accurate approach to expression proteomics. Molecular and Cellular Proteomics, 1 (5): 376–386.

Pandey, A.; Chakraborty, S.; Datta, A.; Chakraborty, N. 2008. Proteomics approach to identify dehydration responsive nuclear proteins from chickpea (*Cicer arietinum* L.). Molecular and Cellular Proteomics, 7: 88–107.

Pechanova, O.; Takáč, T.; Šamaj, J.; Pechan, T. 2013. Maize proteomics: An insight into the biology of an important cereal crop. Proteomics, doi: 10.1002/pmic.201200275.

Pennington, S.R.; Dunn. M.J. 2001. Proteomics: From Protein Sequence to Function. New York: Springer-Verlag and BIOS Scientific Publishers.

Rappsilber, J.; Ryder, U.; Lamond, A. I.; Mann, M. 2002. Large-scale proteomic analysis of the human spliceosome. Genome Research, 12 (8): 1231–1245.

Repetto, O.; Rogniaux, H.; Larré, C.; *et al.* 2012. The seed nuclear proteome. Frontiers in Plant Science, 3 (article 289): 1–7.

Ross, P.L.; Huang, Y.N.; Marchese, J.N.; *et al.* 2004. Multiplexed protein quantitation in Saccharomyces cerevisiae using amine-reactive isobaric tagging reagents. Molecular & Cellular Proteomics, 3 (12):1154–1169.

Scheele, G.A. 1975. Two-dimensional gel analysis of soluble proteins. Characterization of guinea pig exocrine pancreatic proteins. The Journal of Biological Chemistry, Bethesda, 250 (14): 5375–5385.

Schenk, M.F.; Cordewener, J.H.G.; America, A.H.P.; *et al.* 2011. Proteomic analysis of the major birch allergen Bet v 1 predicts allergenicity for 15 birch species. Journal of Proteomics, 74: 1290–1300.

Schulze, W.X. 2010. Proteomics approaches to understand protein phosphorylation in pathway modulation. Current Opinion in Plant Biology, 13: 280–287.

Schulze, W.X.; Usadel, B. 2010. Quantitation in mass-spectrometry-based proteomics. Annual Review of Plant Biology, 61: 491–516.

Schulze, W.X.; Schneider, T.; Starck, S.; *et al.* 2012. Cold acclimation induces changes in *Arabidopsis* tonoplast protein abundance and activity and alters phosphorylation of tonoplast monosaccharide transporters. Plant Journal, 69: 529–541.

Schuster, A.M.; Davies, E. 1983. Ribonucleic acid and protein metabolism in pea epicotyls. II. Response to wounding in aged tissue. Plant Physiology, 73: 817–821.

Schwanhausser, B.; Busse, D.; Li, N.; *et al.* 2011. Global quantification of mammalian gene expression control. Nature, 473 (7347): 337–342.

Serna-Sanz, A.; Parniske, M.; Peck, S.C. 2011. Phosphoproteome analysis of Lotus japonicus roots reveals shared and distinct components of symbiosis and defense. Molecular Plant Microbe Interaction, 24: 932–937.

Shabala, S. 2011. Physiological and cellular aspects of phytotoxicity tolerance in plants: the role of membrane transporters and implications for crop breeding for waterlogging tolerance. New Phytologist, 190: 289–298.

Shen, F.; Hu, Y.; Guan, P.; Ren, X. Ti^{4+}-phosphate functionalized cellulose for phosphopeptides enrichment and its application in rice phosphoproteome analysis. Journal of Chromatography B, 902: 108–115.

Silva, D.I.R. 2012. Caracterização do proteoma nuclear de folhas de cana-de-açúcar (*Saccharum* spp.) de 1 e 4 meses de idade. M.S. Thesis in Genetics and Plant Breeding, ESALQ/University of São Paulo, Piracicaba, SP, Brazil. Available at: http://www.teses.usp.br/teses/disponiveis /11/11137/tde-29112012-144814/ (accessed 26 January 2013).

Silva, J.C.; Gorenstein, M.V.; Li, G.Z.; *et al.* 2006. Absolute quantification of proteins by LCMSE – A virtue of parallel MS acquisition. Molecular and Cellular Proteomics, 5 (1): 144–156.

Simon, G.M.; Cravatt, B.F. 2008. Challenges for the 'chemical-systems' biologist. Natural Chemistry Biology, 4: 639–642.

Soufi, B.; Kumar, C.; Gnad, F.; *et al.* 2010. Stable isotope labeling by amino acids in cell culture (SILAC) applied to quantitative proteomics of *Bacillus subtilis*. Journal of Proteome Research, 7: 3638–3646.

Tanaka, N.; Fujita, M.; Handa, H.; *et al.* 2004. Proteomics of the rice cell: systematic identification of the proteins populations in subcellular compartments. Molecular Genetics and Genomics, 271 (5): 566–576.

Timm, W.; Scherbart, A.; Böcker, S.; *et al.* 2008. Peak intensity prediction in MALDI-TOF mass spectrometry: a machine learning study to support quantitative proteomics. BMC Bioinformatics (9): 443.

Ünlü, M.; Morgan, M.E.; Minden, J.S. 1997. Difference gel electrophoresis: a single gel method for detecting changes in protein extracts. Electrophoresis, 18: 2071–2077.

Vincent, D.; Wheatley, M.D.; Cramer, G.R. 2006. Optimization of protein extraction and solubilization for mature grape berry clusters. Electrophoresis, 27: 1853–1865.

Wang, W.; Scali, M.; Vignani, R.; *et al.* 2003. Protein extraction for two dimensional electrophoresis from olive leaf, a plant tissue containing high levels of interfering compounds. Electrophoresis, 24: 2369–2375.

Wang, W.; Vignani, R.; Scali, M.; Cresti, M. 2006. A universal and rapid protocol for protein extraction from recalcitrant plant tissues for proteomic analysis. Electrophoresis, 27: 2782–2786.

Wang, W.; Tai, F.; Chen, S. 2008. Optimizing protein extraction from plant tissues for enhanced proteomics analysis. Journal of Separation Science, 31: 2032–2039.

Washburn, M.P.; Wolters, D.; Yates, J.R. 2001. Large-scale analysis of the yeast proteome by multidimensional protein identification technology. Nature Biotechnology, 19 (3): 242–247.

Webster, D.E.; Thomas, M.C. 2012. Post-translational modification of plant-made foreign proteins; glycosylation and beyond. Biotechnology Advances, 30: 410–418.

Widjaja, I.; Naumann, K.; Roth, U.; *et al.* 2009. Combining subproteome enrichment and Rubsico depletion enables identification of low abundance proteins differentially regulated during plant defense. Proteomics, 9: 138–147.

Wiese, S.; Reidegeld, K.A.; Meyer, H.E.; Warscheid, B. 2007. Protein labeling by iTRAQ: A new tool for quantitative mass spectrometry in proteome research. Proteomics, 7 (3): 340–350.

Wilkins, M.R.; Sanchez, J.C.; Gooley, A. A.; *et al.* 1995. Progress with proteome projects: why all proteins expressed by a genome should be identified and how to do it. Biotechnology and Genetic Engineering Reviews, 13: 19–50.

Wu, Z.; Moghaddas Gholami, A.; Kuster, B. 2012. Systematic identification of the HSP90 regulated proteome. Molecular and Cellular Proteomics, 11: 1–14.

Ytterberg, A.J.; Jensen, O.N. 2010. Modification-specific proteomics in plant biology. Journal of Proteomics, 73: 2249–2266.

Zhang, B.; Verberkmoes, N.C.; Langston, M.A.; *et al.* 2006. Detecting differential and correlated protein expression in label-free shotgun proteomics. Journal of Proteome Research, 5 (11): 2909–2918.

Zhou, J.Y.; Schepmoes, A.A.; Zhang, X.; *et al.* 2010. Improved LC-MS/MS spectral counting statistics by recovering low-scoring spectra matched to confidently identified peptide sequences. Journal of Proteome Research, 9 (11): 5698–5704.

Zhu, J.; Chen, S.; Alvarez, S.; Asirvatham, V.S.; *et al.* 2006. Cell wall proteome in the maize primary root elongation zone. Extraction and identification of water-soluble and lightly ionically bound proteins. Plant Physiology, 140 (1): 311–325.

Zörb, C.; Herbst, R.; Forreiter, C.; Schubert, S. 2009. Short-term effects of salt exposure on the maize chloroplast protein pattern. Proteomics, 9: 4209–4220.

5　Metabolomics

Valdir Diola (in memoriam),[a] Danilo de Menezes Daloso,[b] and Werner Camargos Antunes[c]

[a]Department of Genetics, Rural Federal University of Rio de Janeiro/UFRRJ, Seropédica, RJ, Brazil

[b]Department of Plant Biology, Federal University of Viçosa, Viçosa, MG, Brazil; and Max-Planck-Institute for Molecular Plant Physiology, Potsdam-Golm, Germany

[c]Department of Biology, Maringá State University/UEM, Maringá, PR, Brazil

Introduction

Plants metabolize more than 200 000 different molecules, which are designed for the structuring and maintenance of tissues and organs as well as the physiological processes associated with growth, development, reproduction, and defense (Weckwerth, 2003). The multiple metabolic pathways are complex, interconnected, and, to some degree, dependent and regulated by their substrates or products. Also, these pathways are influenced by different environmental signals as well as being strongly dependent on the genetic components and the various levels of gene regulation. The central dogma of molecular biology, in the general mode, establishes DNA as a static and structural unit of the genotype, whereas the RNA molecule encodes the protein and is the metabolic functional unit responsible for a specific phenotype. A phenotype can be modulated by a protein (phenotype indirect) and by the presence or amount of one or more metabolites (phenotype direct) at a given time in a tissue. When the phenotype is metabolite-dependent, the variation of these compounds (quantitatively or qualitatively) can be used as a biomarker to identify genotypes which have contrasting accumulations of metabolites of interest. In plants, thousands of compounds are metabolized. The approach that investigate this variation at the metabolite level is called metabolomics, a comprehensive science that interacts with other fields of molecular biology (Figure 5.1).

With the rapid advent of molecular biological techniques and the sequencing of plant genomes in recent years, the understanding of the physiological mechanisms of plants have seen major developments.

Omics in Plant Breeding, First Edition. Edited by Aluízio Borém and Roberto Fritsche-Neto.
© 2014 John Wiley & Sons, Inc. Published 2014 by John Wiley & Sons, Inc.

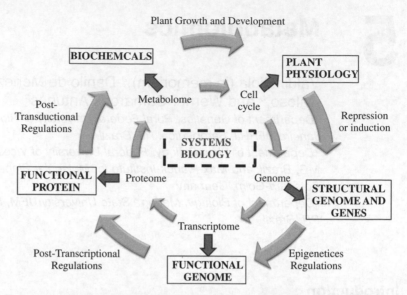

Figure 5.1 Integration between differents levels of control in gene expression. Systems biology is a science whose goals are to discover, understand, and model the dynamic relationships between the biological molecules that build living organisms in order to unravel the control mechanisms inherent to their parts. (See color figure in color plate section).

Systems biology, which includes monitoring of inheritable characters (genomics), the level of transcripts (transcriptome), proteins (proteomics), and metabolites (metabolomic), is of strategic importance in the study of such phenotypes. The investigstion of plant metabolisms is of particular interest due to the wide chemical diversity of plant organisms when compared with animals and microorganisms. Changes in the metabolism may reflect more sensitively the molecular phenotype resulting from a mutation, plant breeding or even a change in the environment.

A particular phenotype can be associated with the contents of specific mRNA and metabolite profiles, enzymes, protein–DNA, protein–RNA, and protein–protein interactions (Kuile and Westerhoff, 2001). The continuous influences of the environment coupled with combined changes in different levels of regulatory molecules are important factors for phenotypic plasticity that is observed in the same genotype or treatment. This is due to the modulation of plant physiological networks under the same hierarchical control (DNA). The simultaneous analysis of more than one regulatory level, such as the association of molecular markers and metabolites, provides a complex database that is useful for plant breeding.

Metabolomic and Biochemical Molecules

The actual technologies that make a qualitative and quantitative overview of metabolites present in an organism possible is called metabolomics (Hall, 2006). The study of plant metabolism provides unique results for improving the understanding of biological information related to the compounds. The size of the metabolome varies, depending on the organism studied. *Saccharomyces cerevisiae*, for example, contains approximately 600 metabolites of the 200 000 so far identified in the plant kingdom (Weckwerth, 2003).

Owing to the high degree of chemical diversity of metabolites found in plants, different techniques are required for the broad investigation of plant metabolism. Various techniques for separation and quantification of such metabolites can be used simultaneously, or in combination, for the measurement of a particular group of metabolites (Dunn and Ellis, 2005). To identify and quantify compounds with differences in molecular weight, polarity, and volatility, it is necessary to use specific methodologies and devices according to the characteristics of each group of compounds (Table 5.1). Therefore, the investigation of a metabolism may involve several analytical platforms that need to be selected carefully according to the metabolites and the metabolic pathway of interest, or the biological question to be adressed (see Chapter 3).

Perhaps the greatest limitation of metabolomics is the low number of metabolites that have so far been identified and quantified reliabily, since there are a large number of metabolites whose chemical structures have not been elucidated. Further, the methodologies available at present show limited capacity for the separation of similar compounds, not allowing identification and quantification to be made. Therefore, the need to increase the list of identified metabolites is a major challenge in metabolomic studies, with the development of the emerging analytical technologies providing more information about metabolism control, contributing greatly to the understanding of functional genomics.

Technologies for Metabolomics

The determination of particular metabolite concentrations is routine in molecular biology laboratories and has been widely used for the study of metabolisms. Several recent terminologies have given new insights into these analytical determinations, such as metabolic profiling, metabolic fingerprinting, and metabolite target profiling, which reflect the differences in coverage, accuracy, and necessary instrumentation for these techniques.

Table 5.1   Technologies commonly used in analysis of plant metabolisms.

Technology[a]	Applications	Properties[b]
GC-MS	The analysis of the lipophilic or polar compounds (e.g., sugars, organic acids, vitamins, tocopherols)	Accuracy: <50 ppm Mass range: <350 Da
GC–GC-MS	Similar to GC-MS, but with a better separation of co-eluting compounds due to increased sensitivity of the combination GC–GC	Accuracy: <50 ppm Mass range: <350 Da
SPME GC-MS	Analysis of volatile compounds (e.g., aromatic compounds, repellents)	Accuracy: <50 ppm Mass range: <350 Da
CE-MS	Analysis of polar compounds (fatty amines, Co-A derivatives, sugars, organic acids, vitamins, tocopherols)	Accuracy: <50 ppm Mass range: <1000 Da
LC-MS	Analysis of secondary metabolites (mainly carotenoids, flavonoids, glucosinolates, vitamins)	Accuracy: 50–100 ppm Mass range: <1500 Da
FT-ICR-MS	High resolution MS in combination with LC is highly potent. Allows the identification of unknown metabolites by the mass/charge ratio (m/z)	Accuracy: <1 ppm Mass range: <1500 Da
NMR	Non-destructive analysis of abundant metabolites in a sample	Accuracy: <1 ppm Mass range: ~50 kDa
Direct injection MS	Non-separative technique that provides a metabolic fingerprint of the content in a biological sample	Accuracy: 50–100 ppm Mass range: <1500 Da
FAIMS-MS	Latest technology for MS. Allows selection of specific ions, reducing ion suppression and matrix effects	Accuracy: 50–100 ppm Mass range: <1500 Da

[a]FT-ICR, Fourier transform ion-cyclotron resonance; FAIMS, field asymmetric waveform ion mobility spectrometry; SPME, solid phase micro extraction.
[b]Da, Dalton.

Although mass spectrometry (MS) is preferrred (Figure 5.2), several techniques can be used to study plant metabolism. In 30% of the main metabolomic techniques, interest has been focused on the identification of biomarkers or metabolites that are associated with a particular genetic or physiological trait (Griffiths *et al.*, 2010) (Figure 5.2). This is due to the success in obtaining data and to the biotechnological techniques. It is noteworthy that, generally, most of the relevant information recorded and published was obtained by using more than one technique, including the latest trends in technology, analytical methods, applications, and biometric data analysis.

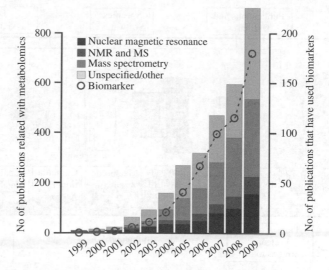

Figure 5.2   Number of papers in PubMed covering the following search terms: metabolomic*, metabonomic*, "metabolic* profiling," "metabolic* fingerprinting," and/or lipidomics (left vertical axis) compared with the following set of terms in the title and in the abstract, as well as the corresponding MeSH keywords: "nuclear magnetic resonance/NMR," "mass spectrometry/MS," and "biological marker/biomarker" (right vertical axis). (Source: Adapted from Griffiths et al., 2010).

Various combinations of analytical techniques for chromatographic separation and identification by mass spectrometry (MS) can be used for the determination of the metabolic profile of a plant sample (see Chapter 8), for example: gas chromatography or high pressure liquid chromatography coupled to mass spectrometry (GC-MS, LC-MS, respectively) and capillary electrophoresis coupled to mass spectrometry (CE-MS) (Lisec et al., 2006). In addition, the technique of nuclear magnetic resonance spectroscopy (NMR) is also commonly used to analyze metabolic profiles, especially in determining the structure of unidentified metabolites, as well as for the analysis of fluxomes (metabolic flux) using labeled molecules, an important component of functional genomics (Fernic et al., 2008). Different combinations of methods have been used in the search for metabolites associated with quantitative trait loci (mQTL).

Although all these techniques have improved substantially, especially since the mid-1990s, the identification and quantification of the levels of all metabolites present in a plant organism are far from being achieved. Furthermore, the identification of metabolites is a challenge that lies not only in obtaining suitable high-quality data, but also in the development of bioinformatics tools for automation of data analysis involving different levels of functional genomics (genome–transcriptome–proteome–metabolome). The integration of data obtained from experimental analysis together with the prospect

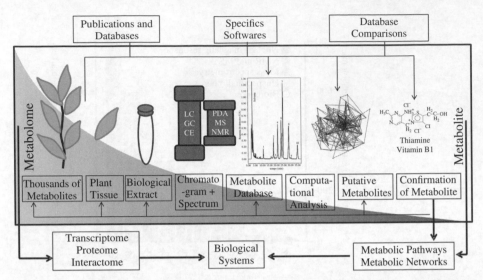

Figure 5.3   Schematic representation of the metabolomics approach to systems biology. The large number of metabolites present in a biological system (e.g., plants) requires a matching design and interpretation of experimental data to identify only a few naturally occurring metabolites in the system that are responsible for specific traits. This procedure includes: sampling, sample preparation and extraction, analysis with capillary electrophoresis (CE), liquid chromatography (LC), gas chromatography (GC), and/or nuclear magnetic resonance (NMR); interpretation of chromatograms and spectra; selection of peaks for statistically relevant candidates; graphical interpretation from multivariate analysis, differential extraction peaks, building a list of putative metabolites and identification of metabolites. During this process, the theoretical resources of species databases, literature, spectral and chemical data can be incorporated into an analysis of a small number of ambiguities between metabolite candidates. The information on the structure and metabolite pool, if associated with the measurements made by dynamic and transient metabolic flux analysis, can help us understand the functioning of metabolic pathways and metabolic networks. (See color figure in color plate section).

of the metabolites being in the database is facilitated by the accessibility of databanks, and the gap between spectrometric signals and biochemical (or chemical) characterization is reduced through some initiatives that help to decrease such differences. The technologies used in analytical and computational metabolomics allow the characterization of molecules through the provision of data that can lead to annotation, and finally to identification (Figure 5.3).

Details of the coupling metabolomics techniques are presented in Figure 5.4 and profile technologies and their applications are described in Table 5.1.

Metabolomic Database Analysis

Before any data analysis, it is worth being aware of the possible sources of variation present in the samples that may influence the final

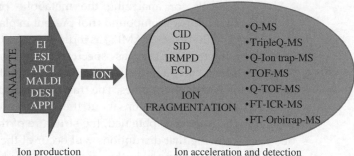

Ion production   Ion acceleration and detection

Figure 5.4 Possible configurations of mass spectrometers. There are various configurations of mass spectrometers according to the acceleration and detection of the ions: quadrupole-MS (Q-MS); triple quadrupole-MS (TripleQ-MS); quadrupole ion trap-MS (Q-ion trap-MS); time-of-flight MS (TOF-MS); Fourier transform MS, ion cyclotron resonance (FT-ICR-MS); and FT-Orbitrap-MS. There are different ionization sources: electron-impact ionization (EI); electrospray ionization (ESI); atmospheric pressure chemical ionization (APCI); matrix assisted laser desorption ionization (MALDI); desorption electrospray ionization (DESI); and atmospheric pressure photoionization (APPI). In terms of technical fragmentation of ions, the most common methods are the collision-induced dissociation (CID), which is the more conventional one. Fragmentation techniques include: surface-induced desorption (SID) and infrared multiphoton dissociation (IRMPD), and also electron capture dissociation (ECD). (See color figure in color plate section).

conclusions, in case they are not being monitored. Parameters such as reproducibility may influence the results (e.g., biological variation between individuals, sampling, sample preparation, and analysis), and they must be monitored, as far as possible by means of measurements with various technical replicates, in order for biological variation to overcome analytical deviation. In principle, the biological variation must overcome all the variance analysis. Details of methods used in the analysis of metabolisms using GC-MS, LC-MS, and NMR can be found in Lisec *et al.* (2006) and Kruger *et al.* (2008).

The identification of the metabolite by MS using a previous chromatographic separation (LC or GC) usually occurs through the combination of exact mass, isotopic distribution, fragmentation patterns, and other information related to the mass/charge ratio (m/z). The results for chemical compounds with a particular m/z must be exactly the same, or very similar, to the library data, which are built-up using ultrapure compounds. This is usually one of the first steps in the identification of the metabolite detected. This set of data will become less complex if the separation and detection methods are selected appropriately, facilitating the identification and quantification of a variety of metabolites. The analysis of a metabolic profile can be pre-selected in accordance with the metabolites available in the library. The library establishes a pattern of retention times in the chromatographic column and an m/z ratio for each metabolite, avoiding errors in the identification of each metabolite.

Many protocols for analyzing the metabolic profile include the addition of a mixture of compounds not present in plants (e.g., an alkane or fatty acid methyl esters, FAMEs) as indicators of retention time, which can help in interpreting the mass spectral data generated. This enables a comparison to be made between different samples running in the same machine at different times. The fragmentation pattern and the m/z ratio can provide information about the structure of the fragmented ion. From the fragments obtained, the structure of the molecule can be deduced, knowing that disruptions will occur at the weakest points of the molecule. For example, in the case of θ-glucosylated flavonoid, the first breakdown occurs on the glycosidic bond and then on the aglucome structure, but only if sufficient energy is provided. This pattern of fragmentation of a single molecule corresponds to fragments having different relative m/z intensities. When a strong form of ionization that produces a constant fragmentation pattern is provided, such as that obtained by GC-EI-MS, there may be large libraries of different fragmentation pattern molecules, which allows the identification of the fragment. The complete profile of fragmentation and the relative intensity between the fragments provides a set of candidates for a single molecule, allowing later unequivocal identification by MS fragmentation.

The unequivocal identification of the compound is an essential step in the quantification and analysis of the metabolic profile. In some cases the identification is difficult due to the low concentration of the compound or to the co-elution of different ions. After identifying the fragmentation profile of a particular compound, the spectra derived from GC-MS can be compared between different databases, as the same column and the same GC-MS parameters will have been used. This enables the establishment and structuring of a database for the analysis of metabolites, such as the platform "Golm metabolome database (GMD)" created by the Max Planck Institute for Molecular Plant Physiology (http://gmd.mpimp-golm.mpg.de). The creation of such a database leads to the standardization of techniques used for analysis of metabolic profiles.

One of the most direct methods to identify an unknown metabolite in a biological sample is to compare it with a commercially available ultrapure standard compound, using the same analytical system. However, this approach requires the availability (commercial) of the ultrapure standard compound, which is rare, especially for secondary metabolisms. In addition, many standard compounds used are unstable and/or show low purity. This can be especially problematic in the MS if the ionization efficiency is relatively high for the impurity. Moreover, it would be hard work to analyze each compound and build it's own database.

In summary, the ability to identify metabolites using MS is limited and lies with the possibility of combining different features of the analysis (exact mass, fragmentation pattern, and isotopic pattern) with other experimental parameters (retention time spectra and UV/Vis), and confirmation with standard compounds. The identification of the "unknown" compounds using these techniques is quite laborious. As for all technologies, "omics" is the multidimensional data characteristic of metabolomics, since the data set is inherently complex.

Multivariate statistical analysis (e.g., principal component analysis (PCA), hierarchical cluster analysis (HCA), partial least squares (PLS), and discriminant analysis (DA)) is widely used in metabolomics. These methods simplify data analysis and through the reduction of dimensionality they may also provide a visual representation of the data.

Sophisticated statistical methods can establish metabolite relationships (e.g., correlation matrices and correlations of metabolic networks) and help establish connections among different metabolites and between metabolites and genes or proteins (Urbanczyk-Wochniak et al., 2005). Thus, an overview of the whole system is obtained. There are different tools for visualization or databases that can be used to display the data from the different coupling "omics" (e.g., KEGG (http://www.genome.jp/kegg/), MetaCyc (http://metacyc.org), MAPMAN (http://mapman.gabipd.org/web/guest), and KappaView (http://kpv.kazusa.or.jp/kpv4/).

Metabolomics Applications

There are a number of potential applications for metabolomics technologies, including the study of metabolic pathways (Saito and Matsuda, 2010), metabolic engineering of plants (Rischer and Oksman, 2006), the secondary metabolism of plants (Keurentjes, 2009), plant nutrition (Keurentjes, 2009), plant development (Peluffo et al., 2010), plant phenotyping (Riedelsheimer et al., 2012), qualitative analysis, biomarkers, and systems biology (Roessner and Beckles, 2009), and others. Moreover, these technologies involve metabolomes that can be used for the detection and identification of biomarkers of diseased plants or plants under abiotic stress, as well as genetic changes. In recent years there have been major advances in genetic engineering, in order to increase, decrease or eliminate the activity of a particular gene and alter its final product (usually a metabolite). Metabolomics, as well as other omic technologies, such as transcriptomics and proteomics, can be used to monitor genetic effects not only under normal growth conditions but also under stress, allowing the estimation and evaluation of the potential risks associated with transgenic and other changes in the genome (Rischer and Oksman-Caldentey, 2006).

Metabolomics is an important tool among the practical genomics techniques, allowing researchers to investigate adaptation mechanisms employed by plants in response to biotic and abiotic stress. The information that is provided by metabolomics offers a better understanding as to what occurs at a biochemical level, contributing to producing new varieties that can withstand adverse conditions, such as drought, salinity, mineral toxicity, and extreme temperatures, without any adverse effects on productivity. A widely known example is that of the osmotic shock response on changing the type and concentration of intracellular solutes. This is usually the result of increased synthesis and accumulation of solutes that act as osmoprotectors, such as polyols (including glycerol and sorbitol), amino acids (especially proline), quaternary nitrogen compounds (glycine betaine), tertiary sulfonic compounds (dimethylsulfoniopropionate), and sugars (Yancey, 2005).

An interesting application of metabolomics is the comparison of the metabolic responses of commercially important plant species to the close parental or wild genotypes (ecotype), which are characterized by high levels of tolerance to a particular stress condition. The metabolites and their pathways may be similarly responsive and they are the basis for adaptation or tolerance of a variety of plant species to stress conditions. The use of this information may result in new breeding strategies, which could lead to the production of new cultivars that grow better under stress conditions and maintain their productivity (Fritsche-Neto and Borém, 2012). Through the analysis of compounds according to their class, Saito and Matsuda (2010) observed changes in metabolism that suggested strategies for stress acclimation. On the other hand, the selection of genotypes through parallel analysis of metabolite fingerprints using LC-MS and GC-MS and molecular marker techniques, enables the detection of metabolites and genomic regions that may present phenotypes of economic interest. In addition, it allows us to understand how populations diverged through evolution over time, an important feature in phylogeny studies (Homuth *et al.* 2012).

Another important application of metabolomics is the elucidation of the metabolic profile changes involving the temporal pattern of organ differentiation. The metabolic maps generated allow inferences about the accumulation of compounds and regulatory mechanisms of development. For example, in a study on the development of coffee beans, it was shown that the accumulation of rare fatty acids and chlorogenic acid in the endosperm occurs during a limited period of the development (Joët *et al.*, 2009). This information is a useful indication of the best phase for the analysis during the search for biomarkers to aid in genetic improvement to the quality of the beverage.

The complexity and plasticity of metabolic networks allow the acclimation and/or adaptations of plants to fluctuating environmental conditions. However, these properties offer a huge obstacle to the metabolic engineering of plants (Zamir, 2008). In this context, metabolomics can contribute significantly to the understanding of the behavior of metabolic networks in the face of various endogenous and/or environmental conditions. Analyses of the metabolic flux profile allow us to explore the dynamics of plasticity of plant metabolism. Under adverse conditions, it becomes a powerful tool for obtaining routes and/or molecules associated with particular genomic regions (QTL) with biotechnological potential.

Metabolomics-assisted Plant Breeding

Metabolomic approaches allow the assessment of a broad range of metabolites and have great value in phenotyping and diagnostic analyzes in plants. These tools are used for the assessment of natural variation, based on the similarity of each metabolic genotype (Fernie and Schauer, 2008). Thus, metabolomics represents an important addition to the tools currently used in genomics and marker-assisted selection in plant breeding. Here, we describe the ongoing progress in the identification of genetic determinants of plant chemical composition, with emphasis on the application of metabolomics strategies and their integration with other technologies.

Plant Breeding for Quality Metabolic Compounds

In recent years, there has been a significant increase in interest in understanding the natural variations in plants. A growing number of research groups have begun studying the complex traits influenced by quantitative trait loci (QTL). Many of these studies concluded that the productivity and resistance of a plant to biotic and abiotic stress are negatively correlated at the molecular level (Fernie and Schauer, 2008). Moreover, the advances achieved with the techniques of the post-genomic era have provided valuable data sets that have helped to clarify associations between genetic variations and plant phenotypes. Although many of these studies were based on the model plant *Arabidopsis thaliana*, the application to crop species is extremely promising (Fernie and Schauer, 2008).

Despite the high costs for the acquisition and maintenance of equipment and materials used in metabolomics, which further limit the use of these tools in some laboratories, a metabolic analysis should be considered as additional information, rather than as an alternative, to plant breeding. However, with the development of new equipment and the emergence of new companies in the field of mass spectrometry, the costs for many methods of post-genomic profiling are rapidly being reduced. The metabolomic profiles have a lower cost than the transcriptomes and do not depend on having pre-genome sequencing available. This eliminates the laborious work on the bioinformatics, which is necessary, for example, in the large-scale analysis of transcriptomes and proteomes. With the latest progress in plant breeding and the demands of the consumer market in recent years, researchers have been focused on the nutritional aspects of food, such as the content and quality of protein, oils, and other compounds, for example, vitamins and antioxidants (Moose *et al.*, 2004, Fernie *et al.*, 2008). Accordingly, the rapid development of high-performance tools for metabolic profiling has facilitated the analysis of a variety of metabolites.

Metabolic engineering in plants, using reverse genetics approaches, often has unexpected consequences, both in the yield of the plants (productivity) and at other levels of the cellular metabolites. This reflects how much of the knowledge gap still needs to be filled in the context of the introduction of metabolomics. The ability to monitor a wide range of metabolites becomes very useful. This allows the detection of desirable characteristics and also facilitates the understanding of the metabolic pathways and how they interact with the observed phenotypes. The objective of most approaches in metabolomics is to analyze chemically complex mixtures of compounds (qualitatively and quantitatively). Some provide a lot of overlapping data (same metabolites) and some are unable to identify all metabolites (unknown) using just a single protocol or platform. Therefore, the potential of metabolomics should be as good as platforms that allow identification/quantification of the maximum number of compounds present in a sample. Given the high chemically diversity of the molecules a single platform run is difficult for all classes of metabolites. Thus we have to make a good choice for each particular case.

Although metabolomics is a new area of science and technology, large amounts of data have been published on the application of the widely divergent genetic populations. These data include assessments of the relative contribution of the genotype and the environment to the composition of metabolites, analysis of heritability, and integration of data on the metabolites with morphological phenotype. A recent study showed that the content of specific metabolites can be enriched in hybrid material by a mechanism that does not lead to a reduction in yield (Riedelsheimer *et al.*, 2012). The main aim of breeding programs is to obtain a particular phenotype, in this case a metabolite-associated phenotype, with no impact on productivity. Along with recent advances in transcription profiling, the integration of data from multiple platforms has become economically viable within a single project. The focus point is the potential of metabolomics in genomics and genetics-assisted breeding where monospecific chemical markers have begun to be frequently generated successfully. The high number of metabolites in a culture may raise technical limitations to teams of scientific researchers, who have traditionally focused on one, or at most, two metabolic characteristics that had the most industrial importance or nutritional value. Prime examples of these specific approaches include the carotenoids content of tomatoes (Fernie *et al.*, 2006), the protein content of maize (Moose *et al.*, 2004; Riedelsheimer *et al.*, 2012), and the starch content of potatoes (Gebhardt *et al.*, 2005).

Metabolites QTLs (mQTLs) Expression

In recent years, researchers have begun to use other approaches to identify genetic determinants for the quality and composition of various plant species. These approaches led to detailed examinations and have increased our understanding of the biosynthesis of glucosinolates (Riedelsheimer *et al.*, 2012), the synthesis of seed oil (Hobbs *et al.*, 2004), oligosaccharides (Bentsink *et al.*, 2000) and flavonoid biosynthesis in *Arabidopsis* (Yonekura-Sakakibara *et al.*, 2007), tomato (Spencer *et al.*, 2005) and *Populus* (Morreel *et al.*, 2006). Moreover, in recent years, several studies have used metabolomic approaches for *Arabidopsis*, tomato, wheat, rice, broccoli, sesame, and mustard (Keurentjes, 2009), which have led to a much deeper understanding of the natural variation of the chemical composition of these species, facilitating the identification of sources of allelic variation that are important in metabolic engineering. Studies on *Arabidopsis* have been performed on three populations of recombinant inbred lines and results showed wide natural variations in primary and secondary metabolisms (Fernie and Schauer, 2008).

The metabolic profile using the LC-MS aproach showed a large variation in quantitative metabolism, and qualitative differences indicated a range of metabolites (Fernie and Schauer, 2008). In addition, this study not only enabled the evaluation of the genetic architecture of the accumulation of aliphatic glucosinolates in *Arabidopsis*, but it also allowed inference of the structure of the underlying pathways. Meyer *et al.* (2007), using GC-MS, analyzed the primary metabolism in recombinant inbred lines (RILs) of *Arabidopsis* var. Columbia. They did not identify any primary metabolites linked with biomass, but did see that indirect metabolic pathways were related to biomass. Other studies using the same population and another introgression line (IL) derived from the same parental line led to the identification of six QTL for metabolic biomass. Two QTLs support the case for the metabolic profile and plant biomass accumulation being correlated. Furthermore, three of the six QTLs could be mathematically predicted purely from the composition of the metabolites. Similar studies were carried out with RIL populations based on two independent experiments for evaluating the heritability of mQTLs compared with those for expressed QTLs (eQTL) determined on the same sample; it was found that mQTLs tend to have lower heritability than eQTLs. Furthermore, statistical analysis of the data set revealed that many mQTLs often exhibit a moderate phenotypic effect, allowing the generation and evaluation of network models that may help to elucidate poorly understood pathways, for example, those involved in the synthesis of volatile compounds and hormones (Fernie and Schauer, 2008).

Simple Studies on the Yield Components

The metabolic profiles of 70 rice cultivars were obtained by Kusano *et al.* (2007) using a combination of 1D and 2D GC-MS, resulting in a database related to nutritional value. In a similar study, although on a smaller scale, Laurentin *et al.* (2008) used a combination of liquid chromatography (HPLC) and AFLP markers to determine the relation between metabolic and genetic diversity in sesame. Interestingly, this study showed that there was a significant difference in diversity patterns from the genomic and metabolic aspects, which indicates that they were not strongly associated with each other. This low metabolome heritability is an initial disadvantage for using metabolomics as a means of selection. The major metabolic plasticity may be more influenced by the environment than the genetic constitution of the individual. This notwithstanding, considering that productivity features with a heritability of about 10% were successfully incorporated into breeding programs, there are expectations that after overcoming the initial obstacles, metabolomics can also be used for selection of traits with low heritability. For example, Wong *et al.* (2004), monitoring the carotenoid content in tomatoes by MALDI-TOF-MS, showed a new system for screening large populations. To this aim, the lines selected from two populations of tomato (*S. pennellii* introgressioned lines and mutants) were analyzed for future applications in breeding for high nutrient levels, showing the importance of metabolomic-assisted breeding to improve the nutritional value of tomatoes.

Studies of Multiple Character Production

A wide variety of compositional characteristics, including protein and oil content, fatty acids, amino acids, organic acid, and sugar contents, were analyzed in three maize hybrids grown at three different locations (Harrigan *et al.*, 2007). Using a similar experimental design, volatile aromatic 2-phenyl-ethanol and 2-phenylacetaldehyde from the tomato were identified after combining metabolic flux profiles with reverse genetics to confirm the biological pathway and association of these aromatic compounds (Tieman *et al.*, 2006). Fernie and Schauer (2008), using GC-MS along with two independent samples, identified 889 QTLs regulating the accumulation of 74 metabolites, including sugars, organic acids, amino acids, and vitamins. This last study revealed that, on average, heritability of mQTLs is associated within an intermediate range (between 0.20 and 0.35) (Fernie and Schauer, 2008). Similar features were observed in corn, which revealed a strong influence of the environment on the metabolite profiles of three genotypes (Harrigan *et al.*, 2007). A comparative study in tomato ILs (isogenic lines) and NILs (near-isogenic lines) showed that most of the mQTLs were inherited dominantly. A considerable number showed additive or recessive inheritance, with only a negligible number displaying overdominant heritage features. The possibility of decoupling the increase in the level of any metabolite without penalties with respect to the performance of crops could be an important advance in the use of metabolomics-associated breeding.

Associative Genome Mapping and mQTL Profiles

Integration of morphological and metabolic profiles may be a promising strategy for plant breeding. Association studies show a strong tendency towards combining metabolomics data with other genomic platforms to provide new perspectives in gene annotation (Hagel *et al.*, 2008) in the regulation of complex biological systems (Homuth *et al.*, 2012). The recent advances in technological platforms have allowed improvements in the capacity of analysis and data processing to be made, providing quantitative analyses of thousands of compounds from a wide variety of chemical classes within a single sample of plant tissue (Lisec *et al.*, 2008).

Clear evidence has recently been presented for the regulation of metabolic networks and their influence on complex traits (Saito and Matsuda, 2010). The plant model *Arabidopsis thaliana* was, and still is, the main source for studies on a variety of metabolites, especially those linked to the accumulation of biomass (Sulpice *et al.*, 2009), illustrating its central role in growth traits and plant development. Metabolomic approaches are being tested in the breeding of different cultures, especially using genetic methodologies to evaluate the natural variability in order to improve the characteristics associated with the chemical composition of agricultural products (Fernie *et al.*, 2006). Some examples of the application of metabolic profiles were conducted to successfully predict productivity in potato (Steinfath *et al.*, 2010) and to distinguish sunflower contrasting genotypes in relation to the response to pathogenic infections (Peluffo *et al.*, 2010). Recently, Riedelsheimer *et al.* (2012) showed that the metabolic profiles of different maize lines (testcross) allowed the prediction of its performance in cultivated fields in different environments. Despite these successes, the genetic basis of the metabolic profiles of the most important crops remains largely unclear. However, certain metabolic products, such as carotenoids in grains and anthocyanins in leaves (Wong *et al.*, 2004), have been genetically well characterized. However, a comprehensive map of the genetic basis with metabolomes is still at the incipient stage. Initial approaches to the study of the genetic basis of the concentrations of many different metabolites were on populations of RILs from *Arabidopsis*, derived from two contrasting parents to map mQTLs (Lisec *et al.*, 2008).

Associative mapping approaches have revealed large numbers of mQTL, but most of them are not able to explain the substantial level of genetic variance (Brotman *et al.*, 2011). However, the association of genetic variation with the concentration of metabolites is interesting for various reasons. Firstly, the mapping of respective genes associated with mQTL can help in the biological function annotation of a metabolite, which may lead to the identification of new biosynthetic pathways. Secondly, new genes that encode enzymes which regulate the metabolic pathways can be identified. Thirdly, mQTL can generate complex functional characteristics (genotype/phenotype associated) for biotechnological or agricultural uses. In crop plants, many important agronomic traits are controlled by a large number of genes with small phenotypic effects. Consequently, using new generation tools with high performance (NGS, Next Generation Sequencing), has become part of the strategy used for Marker-Assisted Selection (MAS). The use of genomic information with thousands of Single Nucleotide Polymorphism (SNPs) is more widely accepted than just QTLs (Riedelsheimer *et al.*, 2012). While this approach will be successful, it provides no biological or mechanistic knowledge of how genetic information is translated into genetic variability of complex traits. Bridging this genotype–phenotype gap remains ones of the majors challenges. A promising approach would, therefore, need to investigate the genetic basis of intermediate phenotypes with reduced genetic complexity, such as production of metabolites, and to link these results with the complex trait of interest (Kim *et al.*, 2011). Although the linkage maps are able to easily dectect one specific QTL to the parental lines, their resolution is very limited due to recombination events and therefore it is inefficient to monitor inheritance in linkage blocks.

With the improvement of new technologies for high throughput genotyping, such as Genome Wide Associations (GWAs), it is possible to dissect quantitative traits. GWA mapping considers the natural Linkage Disequilibrium (LD) generated by recombination events in different ancestral populations. Depending on the level of LD in the population studied, the resolution of the mapping may extend deeper, to the level of a single nucleotide. Although they sound small, these changes in gene level affect the biosynthetic pathway for the production of one or more metabolites and more complex networks of gene regulation. The results obtained so far for *Arabidopsis* are promising and are stimulating further research on other species, as reported by Riedelsheimer *et al.*, (2012) who analyzed the metabolic profile associated with GWAs in maize. These workers studied the association between genetic variants and concentrations of 118 metabolites in the leaves of 289 contrasting corn lines and detected a strong association with 56 110 SNPs. According to this work, the concentrations of metabolites showed a pattern of average correlation of 0.73, consolidating this functional grouping. Most of the

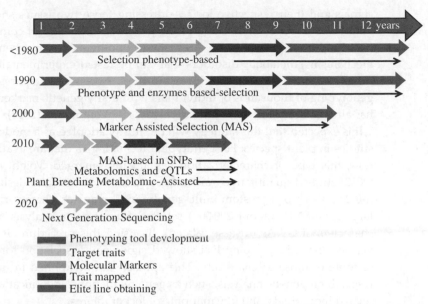

Figure 5.5   The breeding pipeline from 1980 to that envisaged in 2020. In the past, trait discovery was mainly based on phenotypic observations, whereas marker development was restricted to phenotypic or enzymatic or protein markers. Thus, trait mapping and elite line development was a laborious task. The technological advances of molecular biology in the 1980s and 1990s supported the application of molecular markers and improved the speed of trait mapping and commercial material development. Today, the application of MAS in combination with new omics approaches, such as metabolomics or transcriptomics (e.g., eQTL studies) has facilitated the rapid discovery of new traits and allelic variations and, thus, has improved the time to market by several years. In the future, the progress in trait discovery tools, plus simultaneous whole genome sequencing for marker development and trait mapping should shorten the time scale for market introduction of new varieties to 4–5 years. (Source: Adapted from Fernie and Shauer, 2008). (See color figure in color plate section).

observed variances had a strong association with mQTL, especially those related to cell wall synthesis, a target for improving the quality of the biomass.

The new technologies integrated with genetic breeding can be used routinely and MAS should consider the metabolites as a link in narrowing the gap in the genotype–phenotype complex at an agronomic level (Figure 5.5).

Large-scale Phenotyping Using Metabolomics

Recent data from theoretical models and observation of GWAs studies showed that hundreds of loci contribute to complex traits, as was expected. These data raise two main questions. (1) Can additional loci be identified via GWAs variants after meta-analysis? (2) Could research for fine genetic mapping indicate an association between established signals and their respective loci? Addressing these two issues should help us to improve knowledge of the genetic architecture of complex traits and their genetic basis, suggesting hypotheses and biochemical mechanisms/metabolic disorders that can be tested experimentally in functional model systems. Addressing these two issues requires the genotyping of thousands of individuals with many genetic markers, as has already been shown for human diseases by Voight *et al.* (2012).

It is expected that this strategy will be shortly adopted as a model for studies in plant species. For genotyping technologies that are available now, this type of characterization has already emerged. Voight *et al.* (2012) studied quantitative traits related to heart disease and designed the Metabochip, a custom-built genotyping array for characterizing large-scale SNPs (about 200 000) selected from a meta-analysis of 23 characters of GWAs (disease related). By using this procedure, it was verified that 257 mapping loci showed significant association for one or more of those 23 characters. This aproach can be applied in future research on genetic analysis, such as gene regulations, identification of related individuals, and fine mapping of loci of interest.

Strong evidence points to metabolic characteristics with polymorphisms in the genome, and allows the identification of genetic factors as well as observation of specific phenotypes that resulted from its signatures at the metabolome level (Homuth *et al.*, 2012). Similarly, the relationship of transcriptome profiles, metabolome, and proteome data sets produces valuable and revealing associations between the level of the metabolites, mRNA, and the protein. In the future, these promising methods could be applied to plant breeding, providing faster and more accurate results.

Conclusion and Outlook

Clearly metabolomics, despite being a new research area compared with genomics, transcriptomics, and proteomics, will have great impact in the future, as the technologies to detect and identify compounds are improving rapidly, with the development of new equipment and advances with already established platforms. A description of the metabolic profile or the metabolic fingerprint can be achieved through different methodologies, either in parallel or in combination. It can elucidate the differences between treatments and identify metabolites responsible for phenotypic variations. We have highlighted the current status of metabolomics for exploring genetic variation and focused on potential strategies for plant breeding. Although the economic cost and magnitude of heritability must be taken into account, the large body of knowledge accumulated over the years has been in favor of this approach, which should be continuously developed and expanded. The application of post-genomic tools will accelerate the selection and combining metabolomics, genomics, reverse genetics, and large-scale sequencing will probably shorten the time required to produce elite lines. Thus, it is believed that the improvements provided by metabolomics will be widely and routinely applied to species of agronomic interest, providing important, not just complementary, information.

References

Bentsink, L. et al. 2000. Genetic analysis of seed-soluble oligosaccharides in relation to seed storability of Arabidopsis. Plant Physiology, 124: 1595–1604.

Brotman, Y et al. 2011. Identification of enzymatic and regulatory genes of plant metabolism through QTL analysis in Arabidopsis. Journal of Plant Physiology, 168: 1387–1394.

Dunn, W.B.; Ellis, D.I. 2005. Metabolomics: current analytical platforms and methodologies. Trends in Analytical Chemistry, Amsterdam, 24: 285–294.

Fernie, A.R. et al. 2006. Natural genetic variation for improving crop quality. Current Opinion in Plant Biology, 9: 196–202.

Fernie, A.R.; Schauer, N. 2008. Metabolomics-assisted breeding: a viable option for crop improvement? Trends in Genetics, 25 (1): 39–49.

Fritsche-Neto, R.; and Borém, A. 2012. Melhoramento de Plantas Para Condições de Estresses Boticos. Visconde do Rio Branco: Suprema. 240 pp.

Gebhardt, C. et al. 2005. Plant genome analysis: the state of the art. International Review of Cytology, 247: 223–284.

Griffiths, W.G. et al. 2010. Targeted metabolomics for biomarker discovery. Angewandte Chemie International Edition, 49: 5426–5445.

Hagel, J.M. et al. 2008. Quantitative 1H nuclear magnetic resonance metabolite profiling as a functional genomics platform to investigate alkaloid biosynthesis in opium poppy. Plant Physiology, 147: 1805–1821.

Hall, R.D. 2006. Plant metabolomics: from holistic hope, to hype, to hot topic. New Phytologist, 169: 453–468.

Harrigan, G.G. et al. 2007. Impact of genetics and environment on nutritional and metabolite components of maize grain. Journal of Agricultural and Food Chemistry, 55: 6177–6185.

Hobbs, D.H. et al. 2004. Genetic control of storage oil synthesis in seeds of Arabidopsis. Plant Physiology, 136: 3341–3349.

Homuth, G.; Teumer, A.; Völker, U.; Nauck, M. 2012. A description of large-scale metabolomics studies: increasing value by combining metabolomics with genome-wide SNP genotyping and transcriptional profiling. Journal of Endocrinology, 215 (1): 17–28.

Joët, T. et al. 2009. Metabolic pathways in tropical dicotyledonous albuminous seeds: Coffea arabica as a case study. New Phytologist, 182: 146–162.

Keurentjes, J.J.B. 2009. Genetical metabolomics: Closing in on phenotypes. Current Opinion in Plant Biology, 12: 223–230.

Kim, Y.J. *et al.* 2011. Large-scale genome-wide association studies in East Asians identify new genetic loci influencing metabolic traits. Nature Genetics, 43 (10): 990–995.

Kruger, N.J.; Troncoso-Ponce, M.A.; Ratcliffe, R.G. 2008. ^{1}H NMR metabolite fingerprinting and metabolomic analysis of perchloric acid extracts from plant tissues. Nature Protocols, 3: 1001–1012.

Kuile, B.H.; Westerhoff, H.V. 2001. Transcriptome meets metabolome: Hierarchical and metabolic regulation of the glycolytic pathway. FEBS Letters, 500: 169–171.

Kusano, M. *et al.* 2007. Application of a metabolomic method combining one-dimensional and two-dimensional gas chromatography-time-of-flight/mass spectrometry to metabolic phenotyping of natural variants in rice. Journal of Chromatography B, Analytical Technologies in the Biomedical and Life Sciences, 855: 71–79.

Laurentin, H. *et al.* 2008. Relationship between metabolic and genomic diversity in sesame (*Sesamum indicum* L.). BMC Genomics, 9: 250.

Lisec, J.; Schauer, N.; Kopka, J.; Willmitzer, L.; Fernie, A.R. 2006. Gas chromatography mass spectrometry-based metabolite profiling in plants. Nature Protocols, 1: 387–396.

Lisec, J. *et al.* 2008. Identification of metabolic and biomass QTL in *Arabidopsis thaliana* in a parallel analysis of RIL and IL populations. The Plant Journal, 53: 960–972.

Meyer, R.C. *et al.* 2007. The metabolic signature related to high plant growth rate in *Arabidopsis thaliana*. Proceedings of the National Academy of Sciences, U.S.A., 104: 4759–4764.

Moose, S.P. *et al.* 2004. Maize selection passes the century mark: a unique resource for 21st century genomics. Trends in Plant Science, 9: 358–364.

Morreel, K. *et al.* 2006. Genetical metabolomics of flavonoid biosynthesis in *Populus*: a case study. The Plant Journal, 47: 224–237.

Peluffo, L. *et al.* 2010. Metabolic profiles of sunflower genotypes with contrasting response to *Sclerotinia sclerotiorum* infection. Phytochemistry, 71: 70–80.

Riedelsheimer, C. *et al.* 2012. Genomic and metabolic prediction of complex heterotic traits in hybrid maize. Nature Genetics, 44: 217–220.

Riedelsheimer, C. *et al.* 2012. Genome-wide association mapping of leaf metabolic profiles for dissecting complex traits in maize. Proceedings of the National Academy of Sciences U.S.A., 109 (23): 8872–8877.

Rischer, H.; Oksman-Caldentey, K.M. 2006. Unintended effects in genetically modified crops: revealed by metabolomics? Trends in Biotechnology, 24: 102–104.

Roessner U, Beckles DM. (2009). Metabolite measurements. In: Schwender, J. (ed.), Plant Metabolic Networks. New York: Springer, pp. 36–39.

Saito, K.; Matsuda, F. 2010. Metabolomics for functional genomics, systems biology, and biotechnology. Annual Review of Plant Biology, 61: 463–489.

Spencer, J.P. *et al.* 2005. The genotypic variation of the antioxidant potential of different tomato varieties. Free Radical Research, 39: 1005–1016.

Steinfath M, *et al.* 2010. Discovering plant metabolic biomarkers for phenotype prediction using an untargeted approach. Plant Biotechnology Journal, 8: 900–911.

Sulpice, R. *et al*. 2009. Starch as a major integrator in the regulation of plant growth. Proceedings of the National Academy of Sciences U.S.A. 106: 10348–10353.

Tieman, D. *et al*. 2006. Tomato aromatic amino acid decarboxylases participate in synthesis of the flavor volatiles 2-phenylethanol and 2-phenylacetaldehyde. Proceedings of the National Academy of Sciences U.S.A., 103: 8287–8292.

Urbanczyk-Wochniak, E. *et al*. 2005. Profiling of diurnal patterns of metabolite and transcript abundance in potato (*Solanum tuberosum*) leaves. Planta, 221: 891–903.

Voight, B.F. *et al*. 2012. The metabochip, a custom genotyping array for genetic studies of metabolic, cardiovascular, and anthropometric traits. PLOS Genetics, 8 (8): e1002793. doi:10.1371/journal.pgen.1002793.

Weckwerth, W. 2003. Metabolomics in systems biology. Annual Review of Plant Biology. 54: 669–689.

Wong, J.C.; Lambert, R.J.; Wurtzel, E.T.; Rocheford, T.R. 2004. QTL and candidate genes phytoene synthase and zeta-carotene desaturase associated with the accumulation of carotenoids in maize. Theoretical and Applied Genetics, 108: 349–359.

Yancey, P.H. 2005. Organic osmolytes as compatible, metabolic and counteracting cytoprotectants in high osmolarity and other stresses. Journal of Expirmental Biology, 208: 2819–2830.

Yonekura-Sakakibara, K. *et al*. 2007. Identification of a flavonol 7-orhamnosyl-transferase gene determining flavonoid pattern in *Arabidopsis* by transcriptome coexpression analysis and reverse genetics. Journal of Biological Chemistry, 282: 14932–14941.

Zamir, D. 2008. Plant breeders go back to nature. Nature Genetics, 40: 269–270.

6 Physionomics

Frederico Almeida de Jesus, Agustin Zsögön, and
Lázaro Eustáquio Pereira Peres
*Department of Biological Sciences, University of São Paulo/ESALQ,
Piracicaba, SP, Brazil*

Introduction

Plant physiology is a sub-discipline of botany concerned with the functioning of plants. Physiology integrates descriptive knowledge of anatomy and morphology, along with biochemical mechanisms, to study functional processes ranging from seed germination to flowering and fruit set. It is customary to divide plant physiology into sub-disciplines dealing with the environmental factors affecting plants, including: plant–water relations, light and shade, mineral nutrition, and biotic and abiotic stresses, as well as the developmental mechanisms that build the plant body.

A contingency of the sessile lifestyle of plants is that plant development itself is of paramount importance in the response to the environment. Furthermore, plants tend to have a modular form of development, whereby a single module can take the shape of a whole organism or even give rise to one. Thus, post-embryonic development and lifelong organogenesis provide plants with the flexibility to respond adequately to environmental disturbances in spite of lacking locomotive ability.

Plant development is controlled by signaling molecules that have very little in common with their animal counterparts. Some such molecules are plant hormones, historically defined as *naturally occurring, organic substances that influence physiological processes at very low concentrations*. Such signaling molecules participate in the plants' response to the environment, organogenesis, and development in general.

The body of knowledge on plant physiology has accumulated over many decades, and a concise summary of the main discoveries and technical breakthroughs in the area from its inception up to the "omics" era is presented below.

Omics in Plant Breeding, First Edition. Edited by Aluízio Borém and Roberto Fritsche-Neto.
© 2014 John Wiley & Sons, Inc. Published 2014 by John Wiley & Sons, Inc.

Early Studies on Plant Physiology and the Discovery of Photosynthesis

Experimental plant physiology was pioneered by Stephen Hales, whose studies were published in the 1727 book *Vegetable Staticks*. Hales was the first to describe water loss through transpiration, air uptake through leaves and root pressure, along with speculation that plants might use light as an energy source. Further advances took place in the 19th century with the discovery of osmosis by Henri Ducrochet, the description of the nitrogen cycle in soils by Jean-Baptiste Boussingault, and of carbon fixation by Julius von Sachs. The last was not only a major contributor to the field, but also the first person to hold a plant physiology chair named as such in a major university, hence Sachs is usually referred to as "the founding father" of plant physiology (Harvey, 1929).

The 20th century consolidated plant physiology as a scientific discipline and also led to compartmentalization and specialization in the field. Photosynthesis research, for instance, was established and advanced through the work of Otto Warburg starting in 1919. Adapting manometric methods used for gas exchange, Warburg defined a new system for quantum yield studies using the green algae *Chlorella* (Myers, 1974). Gaining widespread acceptance in the research community, *Chlorella* soon became one of the first established plant models. The topic of model plants for plant physiology and physionomics will be discussed later.

Between 1937 and 1940 the biochemist Robert Hill made a major contribution to the understanding of photosynthetic reactions, when he described what is now known as the Hill Reaction (Myers, 1974). This preceded what was probably the greatest breakthrough in photosynthesis research in the 20th century – the discovery of the complete carbon dioxide fixation pathway by Melvin Calvin, Andrew Benson, James Bassham, and co-workers (Bassham *et al.*, 1954). Since these seminal works describing photosynthesis, the greatest advances in plant physiology have taken place in the field of plant development, mainly through the study of plant hormones.

Biochemical Approaches to Plant Physiology and the Discovery of Plant Hormones

Research into plant hormones began early in the 20th century, with the work of the then graduate student Dimitry Neljubow on the triple-response induced by ethylene in peas (*Pisum sativum*) (Saltveit *et al.*, 1998). Further, the substance responsible for the growth of the oat (*Avena sativa*) coleoptile under light, which had been predicted to travel from the tip to the growth region by Charles and Francis Darwin, was physically isolated for the first time and named "auxin" by Dutch botanist Frits Went in 1928. In 1934, Tejiro Yabuta and Yusuke Sumiki discovered the hormone gibberellin (GA) when studying rice (*Oryza sativa*) plants infected by the fungus *Gibberella fujikuroi* (*Fusarium moniliforme*) (Takahashi, 1998).

In the 1950s, research focused on isolating the hormone responsible for cell division. A purinic substance that promoted cytokinesis was discovered by Carlos Miller in 1955, and was named kinetin by Folke Skoog and his group at the University of Wisconsin-Madison (Amasino, 2005). In 1963, Frederick Addicott and Phillip Wareing, working in different groups, isolated abscisic acid (ABA) from cotton (*Gossypium* sp.) and other plant species for the first time.

Early studies of the function of plant hormones utilized two main approaches: exogenous application of the hormone, or an inhibitor; and dosage experiments determining the level of endogenous hormone levels. However, despite their utility, progress in understanding how plant hormones control plant development was hampered by the intrinsic limitations of these approaches. In the case of exogenous applications, the hormone needs to be highly purified, which, depending on its chemical composition, can render it extremely expensive. Furthermore, the ratio between the amount of hormone applied and the amount effectively taken up by the tissue is not known, and neither is the effect of the extraneous hormone on its endogenous counterparts. In the case of inhibitory substances, they have proved useful in some cases, but they are not available for all hormone classes and their specificity is not always high.

High cost is also a major drawback in the hormone dosage approach. The extremely sensitive detection techniques require expensive equipment and reagents. Also, owing to artifacts generated during extraction from the tissue, it is difficult to ascertain whether the chemical species under scrutiny is in fact biologically active. Inactive hormones can therefore be wrongly quantified as active if conjugation or compartmentalization are disrupted during sample preparation, leading to loss of information about the original status of the hormone before extraction.

One of the most successful approaches for the study of hormone physiology has been the use of contrasting genotypes, whether mutant, transgenic or derived from natural genetic variation. The physiology of development was thus boosted during the 1980s, as will be described later.

Genetic Approaches to Plant Physiology and the Discovery of Hormone Signal Transduction Pathways

In the early 1980s the development of molecular biology tools paved the way for the exploration of physiological processes through genetics. Up until then, economically important crops such as maize (*Zea mays*), tomato (*Solanum lycopersicum*), barley (*Hordeum vulgare*), and pea would rarely secure funding for the study of developmental processes (Somerville and Koornneef, 2002). The fruit fly *Drosophila melanogaster* was well established as a genetic model and used by various groups in a highly competitive environment, which precluded any change in research focus at the time (Leonelli, 2007).

There was a need for a new model organism to take advantage of all the recently developed techniques. Thale cress, *Arabidopsis thaliana* (henceforth referred to as "*Arabidopsis*"), an annual plant native to Europe and Asia rose to prominence as a model due to many favorable traits for use in research, such as: small size, short life cycle (four to five weeks), high fecundity (around 10 000 seeds per plant), autogamous reproduction, and ease of crossing allowing generation and conservation of mutations (Meyerowitz and Pruitt, 1985). Collections of artificially generated mutants in *Arabidopsis* had been produced since 1945, so by the early 1980s there was a considerable amount of genetic diversity in the species available for exploration. A large germplasm bank was also available with geographic accessions from a range of different locations; this diversity was later screened for natural genetic variation in flowering time and low-temperature tolerance (Somerville and Koornneef, 2002).

The final push for the widespread adoption of *Arabidopsis* as a genetic model was funding to sequence its genome from the National Science Foundation (NSF) in the mid-1980s. At the time, its nuclear genome (135 million base pairs) was the smallest known in the flowering plants. Its diploid genome, being only five times bigger than that of yeast, allowed for the relatively straightforward assembly of recombinant DNA libraries that facilitated subsequent screening (Meyerowitz and Pruitt, 1985), and development of techniques for genome sequencing (Somerville and Koornneef, 2002).

With ample support from the NSF and a network of *Arabidopsis* research laboratories, the whole-genome sequencing project of *Arabidopsis* was proposed in 1989 (Somerville and Koornneef, 2002). Funding agencies from other countries had also started supporting *Arabidopsis* research and joined the NSF for the sequencing project, thus forming the Arabidopsis Research Initiative in 1990 (Leonelli, 2007).

Sequencing of the *Arabidopsis* genome was completed in the year 2000, covering approximately 120 million base pairs (The Arabidopsis Genome Initiative, 2000). A by-product of the project was the whole-genome array (WGA) from Affymetrix, which has since then been used for differential gene expression analyses, comparative genomic hybridization, chromatin immunoprecipitation, single nucleotide polymorphism (SNP) detection, alternative splicing detection, fusion genes array detection, and Targeting Induced Local Lesions in Genomes (TILLING) arrays (Bevan and Walsh, 2005).

The road to the discovery of gene function directly *in planta* was paved by the use of *Arabidopsis* T-DNA mutant populations together with genome sequence data (Bevan and Walsh, 2005; Krysan *et al.*, 1999). These populations consist of transgenic *Arabidopsis* plants in which a T-DNA insertion interrupts a gene, either knocking it out (producing a null allele if the insert lands within a promoter or coding region) or knocking it down (reducing its expression, when the T-DNA is inserted in an enhancer or 3′ UTR). Such a mutant collection allows for reverse genetic approaches, where the population is screened for mutants in a gene of interest. Genes are thus "tagged" with a T-DNA insertion (T-DNA tagging) and the corresponding mutants can be retrieved via polymerase chain reaction (PCR) using primers specific for both the gene of interest and the T-DNA insert (Krysan *et al.*, 1999).

Understanding of plant development has grown substantially since the completion of the *Arabidopsis* genome sequencing and release of T-DNA mutant populations. The large amounts of information generated on gene sequence and polymorphism (DNA), their expression (RNA), and translation (proteins) were organized in digital databases, which are curated by diverse entities. Most of these resources are publicly available and have become useful tools for the research community. A non-exhaustive list is shown in Tables 6.1 and 6.2.

Plant hormones were one of the areas that benefited the most from the resources generated in *Arabidopsis*. A major breakthrough was the cloning and characterization of the *deetiolated2* (*det2*) gene from *Arabidopsis* (Li *et al.*, 1996). After determining that the gene codes for an enzyme homologous to components of the steroid biosynthesis pathway in animals, it was discovered that brassinosteroids (first isolated and described in the 1970s) are a key plant hormone class involved in essential processes such as cell division and expansion (Clouse and Sasse, 1998).

Even more relevant was the discovery of the long sought-for hormone receptors of the main hormone classes: *ethylene response 1* (*etr1*), which codes for a histidine kinase (Chang *et al.*, 1993); *brassinosteroid-insensitive 1* (*bri1*), a serine-threonine kinase and *cytokinin response 1* (*cre1*), a histidine kinase (Inoue *et al.*, 2001), all of which bind directly to their respective hormone. The ETR1, BRI1, and CRE1 receptors are membrane-bound proteins, which relay the hormone signal through autophosphorylation.

A new signal transduction mechanism was described in 2005, consisting of soluble receptor proteins, involved in the proteosomic

Table 6.1 "Omics" web-based *Arabidopsis* resources (adapted from Lu and Last, 2008).

Database	Description	Site
The Arabidopsis Information Resource (TAIR)	Genome browser and sequence viewer	http://www.arabidopsis.org/cgi-bin/gbrowse/arabidopsis/
Arabidopsis Ensemble Genome Browser (AtEnsembl)	Genome browser	http://atensembl.arabidopsis.info/index.html
Arabidopsis thaliana Integrated Database (ATIDB)	Genome browser	http://www.atidb.org/
Genoscope Arabidopsis Genome Browser	Genome browser	http://www.genoscope.cns.fr/cgi-bin/ggb/arabidopsis/gbrowse/arabidopsis/
Genome View at MIPS *Arabidopsis thaliana* Database (MAtDB)	Genome browser	http://mips.gsf.de/proj/plant/jsf/athal/genomeView/index.jsp
Monsanto Arabidopsis Polymorphism and Ler Sequence Collections at TAIR	Polymorphic molecular markers (SNP and InDel) between Col-0 and Ler accesses	http://www.bar.utoronto.ca/markertracker/
Multiple SNP Query Tool (MSQT)	Polymorphic molecular markers (SNP) between 96 accesses	http://signal.salk.edu/cgi-bin/AtSFP
POLYMORPH	Polymorphic molecular markers (SNP) between 20 accesses	http://borevitzlab.uchicago.edu/resources/molecular-resources
MarkerTracker at the Bio-array Resource for Arabidopsis Functional Genomics (BAR)	RFLP *in silico* analysis and CAPS markers design	http://signal.salk.edu/cgi-bin/tdnaexpress
The SIGnAL Arabidopsis SNP, Deletion and SFP Database	Polymorphic molecular markers (SNP, Del and SFP) between several accesses	http://tilling.fhcrc.org/

Arabidopsis SNP markers from the Borevitz Lab	Polymorphic molecular markers (SNP) between F2 progenies of several accesses	http://www.agrikola.org
T-DNA Express	> 88 000 T-DNA lines created by SALK Institute Genomic Analysis Lab (SIGnAL)	http://signal.salk.edu/cgi-bin/tdnaexpress
Seattle Arabidopsis TILLING Project	Low-cost discovery of induced-point mutations (EMS) for a requested loci	http://tilling.fhcrc.org/
Arabidopsis Genomic RNAi Knock-out Line Analysis (AGRIKOLA)	Seed stock of RNAi-silenced lines	http://www.agrikola.org
Web MicroRNA Designer (WMD)	Applicative for the automated design of artificial microRNAs	http://wmd2.weigelworld.org
AtGenExpress	Tool for visualization of the expression data from AtGenExpress Project	http://www.weigelworld.org/resources/microarray/AtGenExpress/
NASCArrays	Nottingham Arabidopsis Stock Centre's microarray database	http://affymetrix.arabidopsis.info/narrays/experimentbrowse.pl
BAR	Tools for working with functional genomics and other data	http://bar.utoronto.ca/
Genevestigator	Tool for visualization of gene expression data	https://www.genevestigator.com/gv/plant.jsp
Arabidopsis thaliana Co-expression Network (AraGenNet)	Tools for visualization and analysis of transcriptomic co-expression networks	http://aranet.mpimp-golm.mpg.de/aranet
Comprehensive Systems Biology Database (CSB.DB)	Provides results of biostatistical analyses on numeric gene expression data associated with current biological knowledge	http://csbdb.mpimp-golm.mpg.de/
Platform for RIKEN Metabolomics (PRIMe)	Tools for metabolomics, transcriptomics, and integrated analysis of a range of other "-omics" data	http://prime.psc.riken.jp/
CressExpress	Analysis tools for co-expression gene data	http://cressexpress.org/index.jsp

(continued overleaf)

Table 6.1 *(continued)*

Database	Description	Site
A Complete Arabidopsis Transcriptome MicroArray (CATMA)	Transcriptome microarray database	http://www.catma.org/
Arabidopsis Transcriptome Genomic Analysis Database	Transcriptome genomic database	http://signal.salk.edu/cgi-bin/atta
T-DNA Express	Gene mapping tool	http://signal.salk.edu/cgi-bin/tdnaexpress
Arabidopsis thaliana Orphan Transcript Database (AtoRNA DB)	Tool for mapping of ESTs onto the genome sequence, and the subsequent removal of all already annotated transcripts	http://atornadb.bio.uni-potsdam.de/index.php
RIKEN Arabidopsis Full-Length (RAFL) cDNA database	Full-length cDNAs database	http://rarge.gsc.riken.go.jp/cdna/cdna.pl
Salk cDNA and Salk/Stanford/PGEC (SSP) cDNA database	cDNA and EST database	http://signal.salk.edu/2010/index.html
Arabidopsis Massively Parallel Signature Sequencing (MPSS) Plus Database	Small RNAs database, Parallel Analysis of RNA Ends and DNA methylation data generated from bisulfite-treated DNA	http://mpss.udel.edu/#at
Arabidopsis thaliana Small RNA Project (ASRP)	Small RNAs database	http://asrp.danforthcenter.org/
miRNA Precursor Candidates for *Arabidopsis thaliana*	Tool for prediction of miRNAs precursor candidates	http://sundarlab.ucdavis.edu/mirna/
Plant snoRNA Database	Small nucleolar RNAs database	http://bioinf.scri.sari.ac.uk/cgi-bin/plant_snorna/introduction
Cereal small RNAs Database	Small RNAs database	http://sundarlab.ucdavis.edu/smrnas/
SubCellular Proteomic Database (SUBA)	Tool to investigate subcellular localization of proteins	http://www.plantenergy.uwa.edu.au/applications/suba2/
Plant Proteome DataBase (PPDB)	Proteome database	http://ppdb.tc.cornell.edu/
Plastid Protein Database (plprot)	Plastid protein database	http://www.plprot.ethz.ch/

Database	Description	URL
AtProteome Database	Proteome database	http://fgcz-atproteome.unizh.ch/
Arabidopsis 2010 Peroxisomal Protein Database	Peroxisomal protein database	http://www.peroxisome.msu.edu/
Arabidopsis Interaction Viewer at BAR	Protein–protein interaction database	http://bar.utoronto.ca/interactions/cgi-bin/arabidopsis_interactions_viewer.cgi
Arabidopsis thaliana Protein Interactome Database (AtPID)	Protein-protein interaction networks, domain architecture and ortholog information	http://www.megabionet.org/atpid/webfile/
Probabilistic Functional Gene Network of Arabidopsis thaliana (AraNet)	Protein–protein interaction database	http://www.functionalnet.org/aranet/
Membrane-based Interactome Network Database (MIND)	Membrane-based Interactome Network Database	http://cas-biodb.cas.unt.edu/project/mind/index.php
The Golm Metabolome Database (GMD) at CSB.DB	Metabolome Database	http://csbdb.mpimp-golm.mpg.de/csbdb/gmd/gmd.html
KNApSAcK	Species-Metabolite Relationship Database	http://kanaya.naist.jp/KNApSAcK/
Plant Metabolic Network	Metabolic pathway databases	http://www.plantcyc.org/
Kyoto Encyclopedia of Genes and Genomes (KEGG)	Wiring diagrams of molecular interactions, reactions, and relations	http://www.genome.jp/kegg/pathway.html
Pathway Database		
Metabolic Pathway Search at KATANA	Metabolic pathway database	http://www.kazusa.or.jp/katana/pathway.html

Table 6.2   *Arabidopsis* genetic resources (Lu and Last, 2008).

Seed bank	Site
Arabidopsis Biological Resource Center (ABRC)	http://abrc.osu.edu/
Nottingham Arabidopsis Stock Centre (NASC)	http://www.arabidopsis.org.uk/
SALK T-DNA lines	http://signal.salk.edu/cgi-bin/homozygotes.cgi
Lehle Seeds	http://www.arabidopsis.com/
Arabidopsis thaliana Resource Centre for Genomics	http://www-ijpb.versailles.inra.fr/en /plateformes/cra/index.html
CSHL transposon lines	http://genetrap.cshl.org/
GABI-Kat T-DNA lines	http://www.gabi-kat.de/
RIKEN BioResource Center (RIKEN BRC)	http://www.brc.riken.jp/lab/epd/Eng/catalog /seed.shtml

degradation of transcription factor repressors. Specifically, the receptor for auxin *TRANSPORT INHIBITOR RESPONSE 1* (*TIR1*), whose signal transduction corresponds to the repressor degradation model just described, was isolated independently in two laboratories (Dharmasiri *et al.*, 2005; Kepinski and Leyser, 2005). A similar mechanism has been described for the transduction of jasmonic acid (JA) and salicylic acid (SA) signals, which was unveiled when their respective receptor proteins were discovered, CORONATINE-INSENSITIVE PROTEIN 1 (COI1) for JA (Chini *et al.*, 2007; Thines *et al.*, 2007) and NPR1-LIKE PROTEIN 3 and 4 (NPR3, NPR4) for SA (Fu *et al.*, 2012). Another type of signaling, also involving soluble receptors, was described for ABA in 2009, through study of the *pyrabactin resistance 1* (*pyr1*) mutant in *Arabidopsis* (Park *et al.*, 2009).

A notable exception in the hegemony of *Arabidopsis* as a plant hormone model were the discoveries of the GA receptor GIBBERELLIN INSEN-SITIVE DWARF 1 (GID1) (Ueguchi-Tanaka *et al.*, 2005) and of the new hormone class strigolactones (Umehara *et al.*, 2008), both achieved using rice mutants. The use of rice was also pivotal for the discovery of the GA receptor, since the presence of just one copy of the GA-signaling *DELLA* gene (*SLR1*) in this species makes the generation of mutants easier, com-pared with the five copies of *DELLA* genes (*GAI, RGA, RGL1, RGL2,* and *RGL3*) present in *Arabidopsis*. This points to the need for alternative mod-els for the advancement of plant physiology.

Alternative Genetic Models for Omics Approaches in Plant Physiology

Even with the multiple tools available for development studies in *Arabidopsis* (Tables 6.1 and 6.2), evolutionary divergence results in this model not representing the whole diversity of physiological processes in angiosperms. Alternative models have been developed to fill in the gaps not covered by *Arabidopsis*, such as rhyzobial and mycorrhizal interactions, CAM and C4 photosynthesis, sympodial flowering, tuber development, compound leaf development, glandular trichome formation, and fleshy fruit development, among others (Campos *et al.*, 2010; Carvalho *et al.*, 2011; Krysan *et al.*, 1999). Owing to the considerable reduction in genome sequencing costs, many other plant species used as genetic models have also had their entire genome sequenced and other sequencing projects are currently underway.

Throughout their evolution, leguminous plants (Fabaceae) developed mechanisms of interaction with soil microorganisms, leading to symbiotic relationships with nitrogen-fixing bacteria unique among flowering plants. Biological fixation of nitrogen has been studied in soybean (*Glycine max*), *Medicago truncatula*, and *Lotus japonicus* (Libault *et al.*, 2010; Panara *et al.*, 2012). Complete genome sequences are available for these three species. In the case of soybean, sequencing contributed to unveiling the genetic basis of yield-related traits, thus leading to the development of new elite varieties (Schmutz *et al.*, 2010). A considerable drawback, however, is the polyploid nature of the soybean genome, with at least 75% of its genes present in multiple copies (Schmutz *et al.*, 2010). This renders soybean impractical as a model for studies using reverse genetic approaches. The smaller and diploid genomes of *M. truncatula* and *L. japonicus*, on the other hand, have contributed to their widespread use as models for the study of rhyzobial (nitrogen fixation by *Rhyzobium* spp.) and mycorrhizal (phosphorus uptake) interactions. Mutants have been generated in these species using insertion (T-DNA and transposon), chemical (EMS), and physical (gamma-ray) methods, which has led to the isolation of specific genes involved in the respective symbioses (Thoquet *et al.*, 2002). The concomitant use of both species as models is justified in their being phylogenetically distant and thus having different systems of nodulation.

The tomato has long been exploited as a model for the development of fleshy fruits (Emmanuel and Levy, 2002). It also complements *Arabidopsis* as a model for the study of glandular trichomes (Campos *et al.*, 2009), sympodial flowering (Pnueli *et al.*, 1998), compound leaf development (Hareven *et al.*, 1996), and mycorrhizal association (Zsögön *et al.*, 2008). Tomato shares many of the advantages of *Arabidopsis* as a model species, being diploid, autogamous, and harboring a relatively small genome (900 Mb), which has recently been fully sequenced (The Tomato Genome Consortium, 2012). The diploid nature and reduced chromosome complement ($n = 12$) of tomato has allowed the generation of a plethora of monogenic mutants (Rick, 1986). Various resources are publicly available for tomato, including saturated genetic maps with numerous morphological and molecular markers associated with traits of biological

Table 6.3   "Omics" web-based tomato resources.

Data base	Description	Site
Sol Genomics Network	Genome browser	http://solgenomics.net /organism/Solanum _lycopersicum/genome
Tomato Functional Genomics Database	Database for RNA-seq and microarray, metabolite/gene expression correlation tool, small RNA searching tool, and miRNA target prediction tool	ted.bti.cornell.edu/
Tomatoma	Database for full-length cDNA and BAC sequences of the tomato cultivar Micro-Tom	http://tomatoma.nbrp .jp/index.jsp
KaTomics DB	Databases for DNA markers, SNP annotations, and genome sequences of the tomato cultivars Ailsa-Craig, M82, and Micro-Tom	http://www.kazusa.or .jp/tomato/
Real Time QTL	*In silico* genetic components of tomato QTLs	http://zamir.sgn.cornell .edu/Qtl/Html/home.htm
Tomato Epigenome Database	Database for DNA methylation, small RNAs and gene expression profiles of tomato fruits at four developmental stages	http://ted.bti.cornell .edu/epigenome/

Table 6.4   Tomato seed stock resources.

Seed bank	Site
Tomato Genetic Resource Center (TGRC)	http://tgrc.ucdavis.edu/
The Genes That Make Tomato	http://zamir.sgn.cornell.edu/mutants/links /abstract.html
Tomatoma	http://tomatoma.nbrp.jp/index.jsp
Lico TILL	http://www.agrobios.it/tilling/index.html
HCPD Lab Micro-Tom Mutants	http://www.esalq.usp.br/tomato/

and economic importance (Table 6.3). A considerable number of mutants have been well characterized and are available in open germplasm collections (Table 6.4).

On top of induced mutations, natural genetic variation is found in abundance in wild relatives of tomato in the *Solanum* genus, section *Lycopersicon*, most of which can be readily crossed to tomato (Taylor, 1986; Stevens and Rick, 1986). These species evolved in a geographic area with considerable latitudinal range, from southern Ecuador to northern Chile (Warnock, 1991). Within this area, arid, coastal and elevated

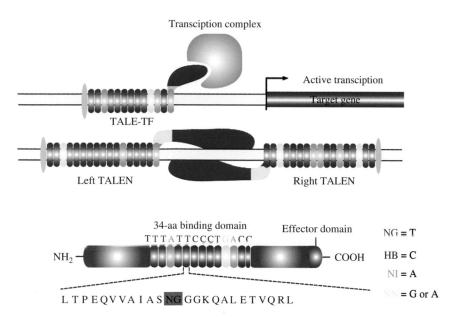

Figure 1.4 Transcription activator-like effector nucleases (TALENs).

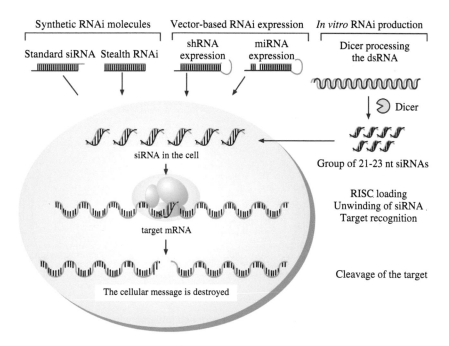

Figure 1.5 Pathways of gene silencing in plant cells. (Source: Based on Souza *et al.*, 2007).

Omics in Plant Breeding, First Edition. Edited by Aluízio Borém and Roberto Fritsche-Neto.
© 2014 John Wiley & Sons, Inc. Published 2014 by John Wiley & Sons, Inc.

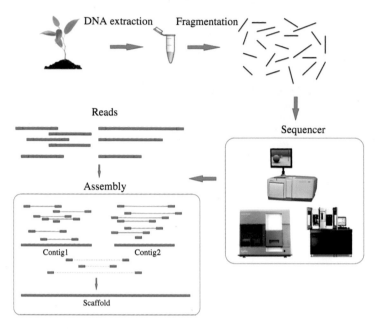

Figure 2.1 Basic scheme, representing the process of plant DNA sequencing.

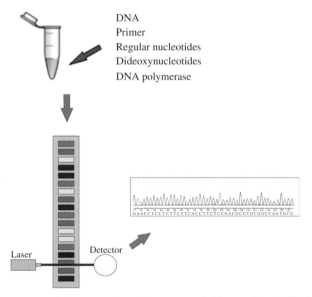

Figure 2.2 Sanger automatic sequencing. The process initiates with a PCR in which, besides DNA, primers and deoxynucleotides, dideoxyribonucleotides (ddNTP) are added, which are labeled with fluorescent markers and without hydroxils at the 3ʹ-position. Thus, every time a ddNTP is added to the chain, its extension is interrupted and, after numerous cycles, many different size fragments will be generated, which have been terminated by different ddNTPs. These fragments are later subjected to electrophoresis and separated into their different sizes. The fragments pass through a laser and the fluorescence is detected; this signal is then transformed into a chromatogram (a graph with peaks of different colors), each colored peak being attributed to a different base. The sequencer comes with software that will convert chromatograms into FASTA format sequences.

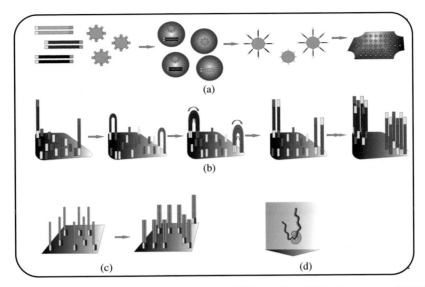

Figure 2.3   NGS clone amplification: (a) emulsion PCR, used by 454, Polonator, and ABI SOLID platforms; (b) bridge amplification used by Illumina; (c) single molecule sequencing used by Helicos; and (d) real time sequencing, used by SMRT. (Source: Adapted from Metzker, 2009; Shendure and Ji, 2008).

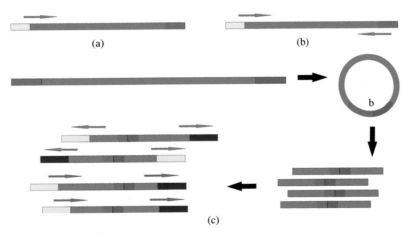

Figure 2.4   Reads generated by NGS. (a) Single-end reads, where the reading is performed only on one end (indicated by the blue arrow). (b) Paired-end reads, where the sequencing is performed in both fragment ends, in opposite directions (blue arrows). (c) In the mate-pair reads, biotin-labeled nucleotides are linked to both ends of the fragment, which is circularized and cut in small sequences that are selected by biotin-based rescue and moved to sequence reading (single by 454 and paired by Illumina and SOLID. (Source: Adapted from Hamilton and Buell, 2012).

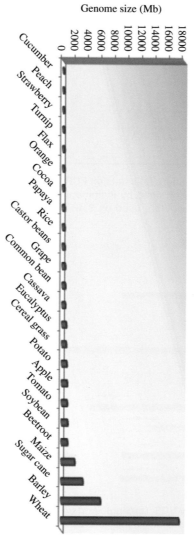

Figure 2.5   Genome size of cultivated species. Published data (gray) and four economically important species (orange).

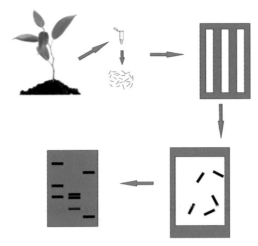

Figure 2.6 Scheme representing RFLP. The DNA is extracted and fragmented and the fragments are separated by electrophoresis in agarose gels, where a continuous smear is seen. The fragments are denatured and transferred to nylon or nitrocellulose membranes, followed by hybridization with a radioactive fluorescent probe containing a single-strand target fragment. Autoradiography, in which the membrane hybridized with the labeled probe is exposed to an X-ray film.

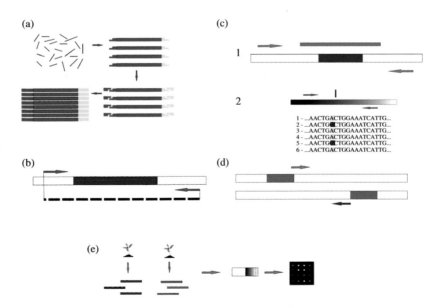

Figure 2.7 Molecular marker depictions: (a) AFLP; (b) SSR; (c) SNP; (d) ISSR; and (e) DArT.

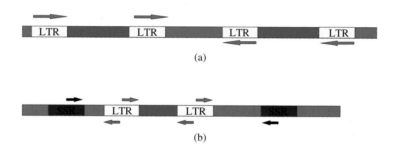

Figure 2.8   IRAP (a) and REMAP (b) marker techniques.

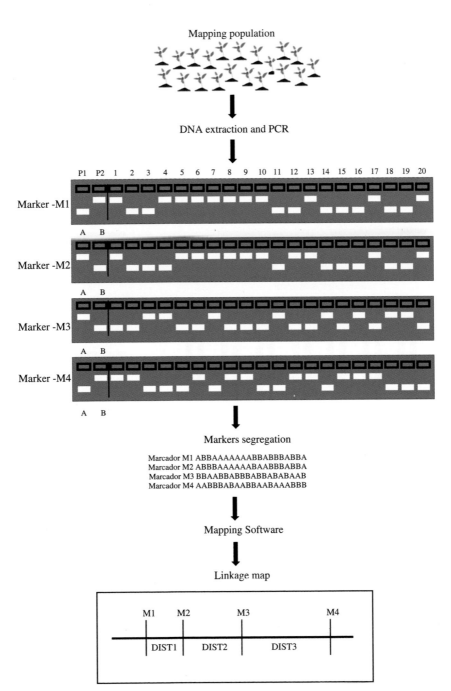

Figure 2.9   Linkage map based on a small population (20 individuals) of recombination inbred lines, based on the segregation of molecular markers. The map is generated by software that analyzes the probabilities of marker orders and distances between them, attaining the best fit (Source: Adapted from Collard *et al.*, 2005).

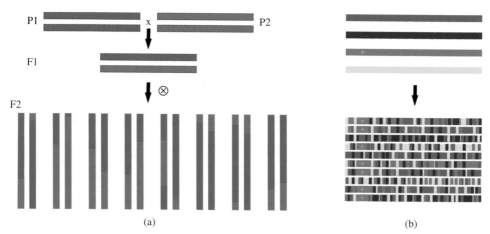

(a)   (b)

Figure 2.10   Comparison of mapping strategies. (a) Biparental mapping is based on populations originating from controlled crosses (the example is showing an F^2 population), with few
recombination opportunities and large extensions suffering from linkage desequilibrium. (b)
Association mapping is performed on populations that have undergone many generations and
therefore, many opportunities for recombination. Also, there are small extensions of linkage
desequilibrium. The white star represents a cluster of linked genes, showing how rare the occurrence of recombinations is when the loci are very close (Source: Adapted from Zhu *et al.*, 2008).

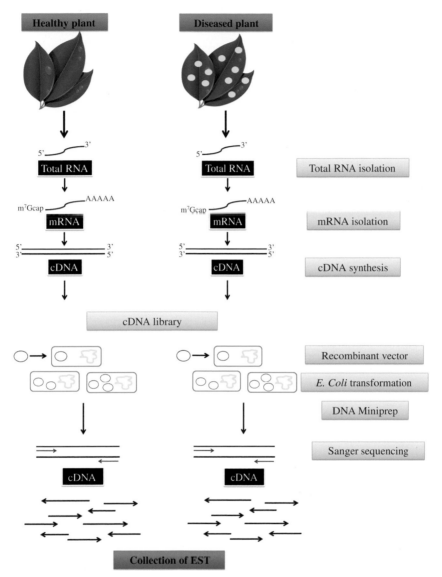

Figure 3.1 Overview of cDNA construction and ESTs generation. The process starts with the isolation of mRNA from the total RNA individually isolated from samples of interest (in this example, leaf samples of diseased and healthy/control plants). mRNAs are reverse transcribed to produce complementary DNA (cDNAs) creating libraries of cDNAs cloned into an appropriate vector. After cloning and *E. coli* transformation, individual clones from these libraries are randomly selected and subjected to a single sequencing reaction using universal primers; one or both ends of the cloned fragment (insert) are sequenced. The collection of short fragments (ESTs) generated is then processed using a number of bioinformatics tools. In this example, it is also possible to use the relative abundance of ESTs representing the same gene to compare the level of gene expression between two experimental conditions.

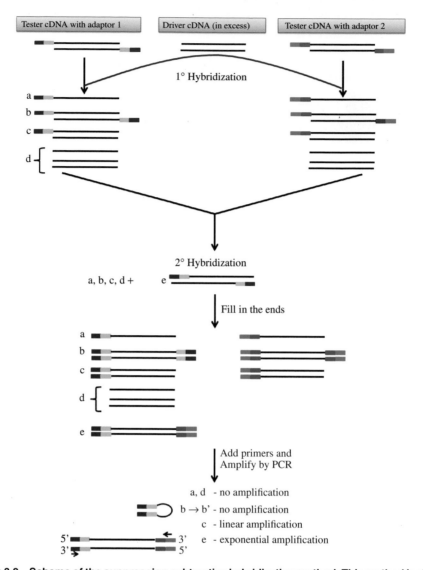

Figure 3.2 Scheme of the suppression subtractive hybridization method. This method includes several steps. Firstly, cDNAs from tester and driver samples are synthesized and digested with a restriction enzyme. Adapters (1 and 2) are ligated to one of two aliquots of tester sample. The first hybridization is performed mixing an excess of driver with both aliquots of the tester. In this step, the subset of tester molecules is normalized: more abundant cDNA fragments anneal faster forming double-stranded (ds) homohybrids (type "b"), while less abundant cDNA fragments remain single-stranded (ss) (type "a"). Also, complementary fragments originating from the same cDNA anneal, resulting in tester-driver double-stranded heterohybrids (type "c"). Secondly, the two samples from the first hybridization are pooled and more freshly denatured driver is added. During the second hybridization, only ss-molecules (type "a") are able to reassociate and form type "b", "c", and "e" hybrids. The least is the tester–tester heterohybrid, having a different adapter at each end. After the second hybridization, end-filling and PCR, "a" and "d" molecules lack primer annealing sites and cannot be amplified; "b" molecules form stem–loop structures that suppress amplification (suppressive effect); "c" molecules are linearly amplified because they have only one primer annealing site. Type "e" molecules are exponentially amplified using primers that anneal at the two different primer annealing sites. Scheme according to the instruction manual (PCR- Select™ cDNA subtraction kit, Clontech).

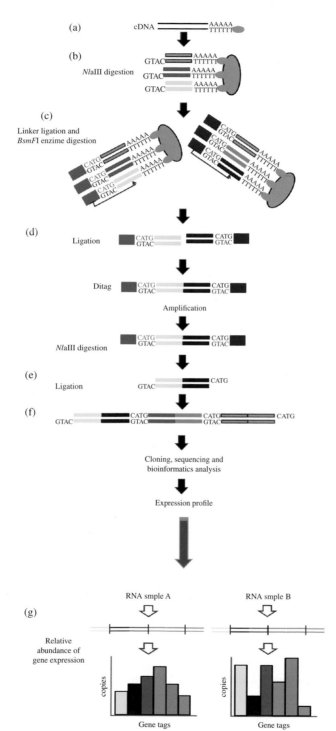

(a) cDNA AAAAA TTTTTT

(b) Nla III digestion
GTAC AAAAA TTTTTT
GTAC AAAAA TTTTTT
GTAC AAAAA TTTTTT

(c) Linker ligation and BsmFI enzime digestion
CATG GTAC AAAAA TTTTTT
CATG GTAC AAAAA TTTTTT
CATG GTAC AAAAA TTTTTT
CATG GTAC AAAAA TTTTTT
CATG GTAC AAAAA TTTTTT
CATG GTAC AAAAA TTTTTT

(d) Ligation
CATG GTAC   CATG GTAC

Ditag
CATG GTAC   CATG GTAC

Amplification

Nla III digestion
CATG GTAC   CATG GTAC

(e) Ligation
GTAC   CATG

(f)
GTAC   CATG GTAC   CATG GTAC   CATG

Cloning, sequencing and bioinformatics analysis

Expression profile

(g)
RNA smple A   RNA smple B

Relative abundance of gene expression

copies

Gene tags   copies   Gene tags

Figure 3.3 Schematic representation of serial analysis of gene expression (SAGE). (a) cDNA synthesis using reverse transcriptase and oligo-dT primers attached to magnetic beads. (b) cDNAs are digested with the restriction endonuclease Nla III. (c) Adapters are ligated to the digested cDNA. These adapters have a recognition site for the enzyme BsmF I, which cleaves at a fixed distance downstream from its recognition site. Cleavage releases the sequences from the magnetic beads. (d) Each released cDNA tag is connected to another forming a ditag. (e) Ditags are amplified by PCR using primers that anneal to the adapters. Subsequently, the amplified fragments are cleaved with Nla III to release the adapters. (f) Ditags are ligated to form a con-catemer, which is then cloned into vectors and sequenced, generating the SAGE library. (g) The number of tags found within the same sequence permits us to deduce the relative abundance of gene expression. (Source: Diagram from Garnis et al., 2004).

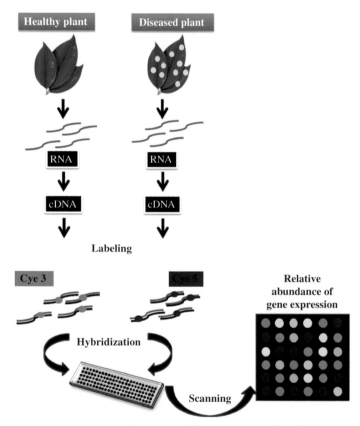

Figure 3.4 Analysis of gene expression using a DNA microarray. Total RNA is isolated from two different samples (healthy and diseased plant), purified, and used as templates for the synthesis of cDNAs, which are labeled with different fluorescent (Cye 3 or Cye 5) dyes for each condition. The labeled cDNAs are mixed and hybridized against the probes (DNA/cDNA known) immobi-lized on the microarray chip. Laser excitation of the fluorophores produces an emission with a specific spectrum, which is captured by a scanner and analyzed by software. The resulting color intensity is associated with the emitted fluorescence from each sample hybridized, which is proportional to the level of gene expression.

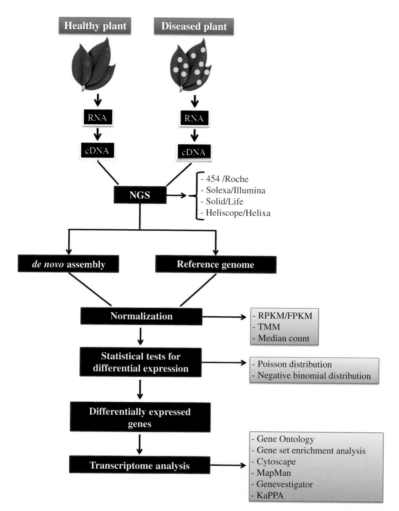

Figure 3.5 Schematic representation of gene expression analysis using RNA-seq. RNA is isolated from two contrasting samples (healthy and diseased plant), and used for the synthesis of cDNAs. These are sequenced using one of the NGS methods, generating millions of reads for each sample. The reads are mapped against a reference genome or *de novo* assembly is performed. After this step, the data are normalized, and then statistical tests are performed to identify differentially expressed genes. Certain programs can infer the biological functions of these genes.

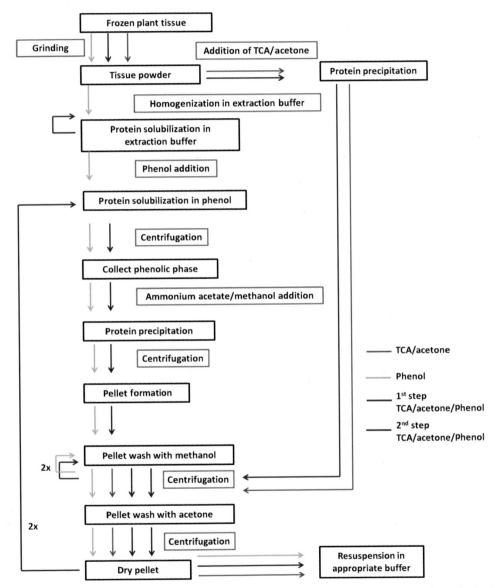

Figure 4.1 Diagram illustrating the three predominant methods for the extraction of plant proteins.

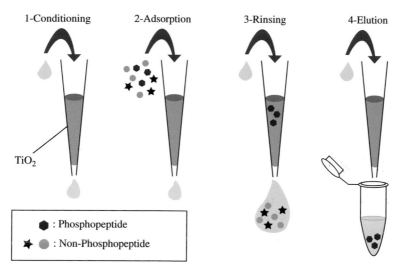

1-Conditioning   **2-Adsorption**   **3-Rinsing**   **4-Elution**

TiO_2

● : Phosphopeptide

★ ● : Non-Phosphopeptide

Figure 4.2 Enrichment of phosphopeptides through TiO² columns, showing the different steps of the procedure. Modified from "Using GL Sciences' Titansphere® Phos-TiO Kit" (www.glsciencesinc.com).

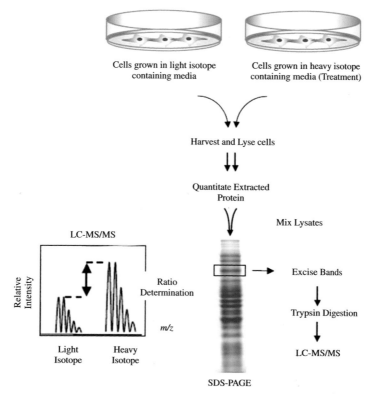

Cells grown in light isotope containing media

Cells grown in heavy isotope containing media (Treatment)

Harvest and Lyse cells

Quantitate Extracted Protein

Mix Lysates

LC-MS/MS

Relative Intensity

Ratio Determination

m/z

Light Isotope   Heavy Isotope

Excise Bands

Trypsin Digestion

LC-MS/MS

SDS-PAGE

Figure 4.3   Schematic of SILAC Workflow, adapted from Thermo Scientific Pierce catalog for Mass Spectrometry Sample Preparation–V2, 2010.

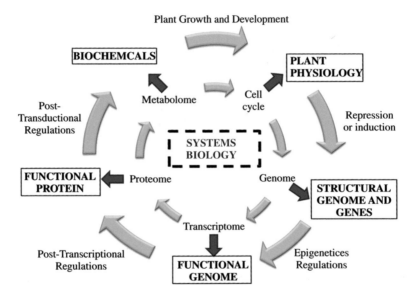

Figure 5.1 Integration between differents levels of control in gene expression. Systems biology is a science whose goals are to discover, understand, and model the dynamic relationships between the biological molecules that build living organisms in order to unravel the control mechanisms inherent to their parts.

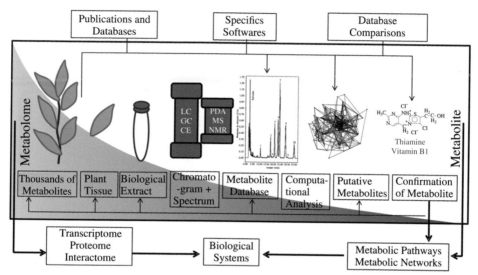

Figure 5.3 Schematic representation of the metabolomics approach to systems biology. The large number of metabolites present in a biological system (e.g., plants) requires a matching design and interpretation of experimental data to identify only a few naturally occurring metabolites in the system that are responsible for specific traits. This procedure includes: sampling, sample preparation and extraction, analysis with capillary electrophoresis (CE), liquid chromatography (LC), gas chromatography (GC), and/or nuclear magnetic resonance (NMR); interpretation of chromatograms and spectra; selection of peaks for statistically relevant candidates; graphical interpretation from multivariate analysis, differential extraction peaks, building a list of putative metabolites and identification of metabolites. During this process, the theoretical resources of species databases, literature, spectral and chemical data can be incorporated into an analysis of a small number of ambiguities between metabolite candidates. The information on the structure and metabolite pool, if associated with the measurements made by dynamic and transient metabolic flux analysis, can help us understand the functioning of metabolic pathways and metabolic networks.

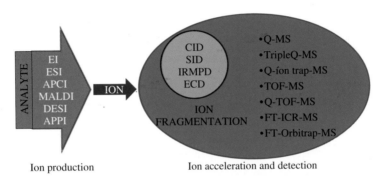

Ion production   Ion acceleration and detection

**Figure 5.4 Possible configurations of mass spectrometers. There are various configura-
tions of mass spectrometers according to the acceleration and detection of the ions:
quadrupole-MS (Q-MS); triple quadrupole-MS (TripleQ-MS); quadrupole ion trap-MS (Q-ion
trap-MS); time-of-flight MS (TOF-MS); Fourier transform MS, ion cyclotron resonance
(FT-ICR-MS); and FT-Orbitrap-MS. There are different ionization sources: electron-impact ionization (EI); electrospray ionization (ESI); atmospheric pressure chemical ionization (APCI); matrix
assisted laser desorption ionization (MALDI); desorption electrospray ionization (DESI); and
atmospheric pressure photoionization (APPI). In terms of technical fragmentation of ions, the
most common methods are the collision-induced dissociation (CID), which is the more conventional one. Fragmentation techniques include: surface-induced desorption (SID) and infrared
multiphoton dissociation (IRMPD), and also electron capture dissociation (ECD).**

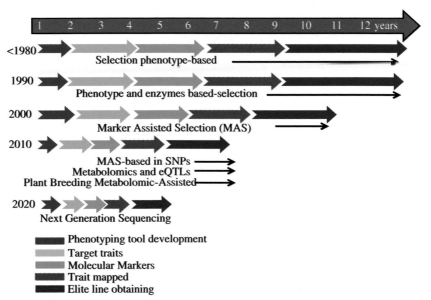

Figure 5.5 The breeding pipeline from 1980 to that envisaged in 2020. In the past, trait discovery was mainly based on phenotypic observations, whereas marker development was restricted to phenotypic or enzymatic or protein markers. Thus, trait mapping and elite line development was a laborious task. The technological advances of molecular biology in the 1980s and 1990s supported the application of molecular markers and improved the speed of trait mapping and commercial material development. Today, the application of MAS in combination with new omics approaches, such as metabolomics or transcriptomics (e.g., eQTL studies) has facilitated the rapid discovery of new traits and allelic variations and, thus, has improved the time to market by several years. In the future, the progress in trait discovery tools, plus simultaneous whole genome sequencing for marker development and trait mapping should shorten the time scale for market introduction of new varieties to 4–5 years. (Source: Adapted from Fernie and Shauer, 2008).

Figure 7.2 IRGA – Infrared Gas Analyzer.

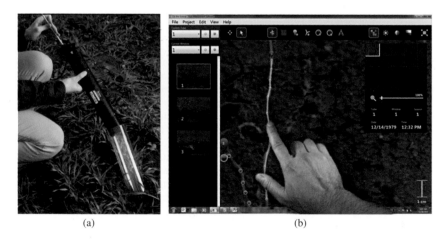

(a)   (b)

Figure 7.3 Scanner for collection of root images (a) and *output* of RootSnap software (b).

Figure 7.4 Image collection of an experiment. (Source: www.lemnatec.de. Reproduced with permission of LemnaTec GmbH, Aachen, Germany).

Figure 7.5 Infrared thermal image of a genotype under stress condition, showing temperature distribution along measured plant. (Source: www.lemnatec.de. Reproduced with permission of LemnaTec GmbH, Aachen, Germany).

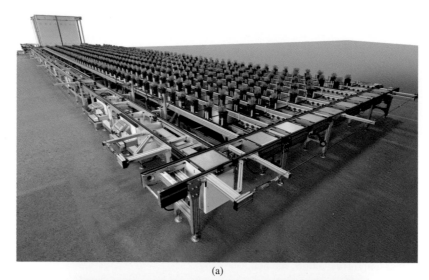

(a)

(b)

Figure 7.6 Scanalyzer^{3D} Discovery Platform. Plants are taken individually by conveyor belt to the image collection cabin (a), where a digital image is collected from multiple angles (b). (Source: www.lemnatec.de. Reproduced with permission of LemnaTec GmbH, Aachen, Germany).

Figure 7.8 Near-infrared spectrophotometry sensors installed on tractors for large-scale phenotyping.

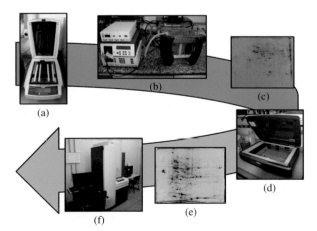

Figure 8.1 (a) Isoelectric focusing system; (b) electrophoresis system; (c) 2DE gel stained with Coomassie blue; (d) 2D gel image scanner; (e) 2D gel digital image; and (f) mass spectrometer.

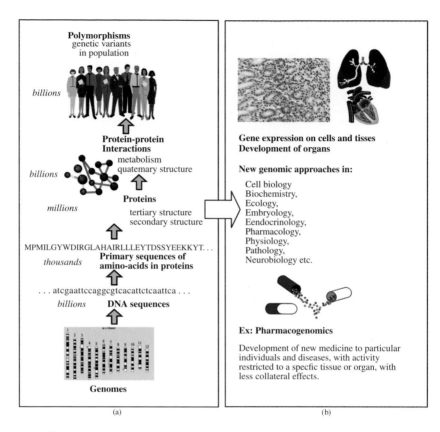

Figure 9.1 Biological data increase (a) and practical applications of the genomic knowledge (b).

```
(a) FASTA
>B07.esd    1280       31     125  ESD
ACTATAGGGCACACTTCCGTACGAAGATACTGGGTACGCGTAAGCTTGGG
CCCCTCGAGGGATACTCTAGAGCGGCCGCCCTTTTTTTTTTTTGGTGGTTT
TTTTTTTTTTTTTTTTTTACTTTACAAAATCTTCGTTTCTTCTATTGTGAACT
ATGATATGATCACGTATAGCAGTAAATCTTGCGGATAGTTGACGAAGTAA

(b) FASTA.qual
>B07.esd    1280       31     125  ESD
 0  0  0  0  0  0  0  0  0  0  0  0  0  0  0  0  0  0  0
 0  0  0  0  0  0  0  0  0  0  0  0  0  0 11 11 20
38 20 20 21 21 23 24 35 31 33 38 27 27 27 27 27 29
32 32 32 32 38 27 27 27 27 27 27 38 37 29 27 31
31 31 31 37 31 31 27 20 20 20 20 20 21 27 26 26 28
33 33 33 26 18 13  8   8  10 10 10 10 18 23 30 33
33 28 28 19 19 19 21 21 21 21 25 22 24 24 22 22 14
12  8  9 14 14 18 17 17 13 16 15 21 13 13 19 24 18
25 18 14 12 11  8   7  7  7  9 13 23 25 23 23 18 15
11 10  0  0  0  0  0  0  0  0  0  0  0  0  0  0  0
 0  0  0  0  0  0  0  0  0  0  0  0  0  0  0  0  0
```

```
(c) FASTQ
@B07.esd
ACTATAGGGCACACTTCCGTACGAAGATACTGGGTACGCGTAAGCTTGGG
CCCCTCGAGGGATACTCTAGAGCGGCCGCCCTTTTTTTTTTTTTGGTGGTTT
TTTTTTTTTTTTTTTTTTACTTTACAAAATCTTCGTTTCTTCTATTGTGAACT
ATGATATGATCACGTATAGCAGTAAATCTTGCGGATAGTTGACGAAGTAA
+
!!!!!!!!!!!!!!!!!!!!!!!!!!!!!!!!!!!!,,5G556689D@BG<<<<
>AAAAG<<<<<<<GF><@@@@F@@<555556<;;=BBB;3.))+++++38
?BB==4446666:79977/)**//322.106..493:3/,)(((*.8:88
30,+!!!!!!!!!!!!!!!!!!!!!!!!!!!!!!!!!!!!!!!!!!!!!!!!!!
```

Figure 9.2 DNA sequence in FASTA format (a) and quality file FASTA.qual (b), both generated by *Phred* software, compared with *FASTQ* format (c). *FASTQ* files incorporate the sequence of bases identified by the "@," separated by the symbol "+" from quality data encoded by different characters, for example, the characters "." "+," and "5" match the *Phred* values 0, 10, and 20, respectively. Regions of low quality (bases in red) had the quality values set to zero and will be dropped from the processed sequence.

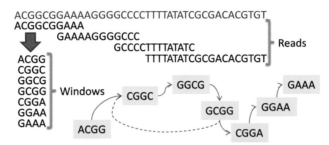

Figure 9.3 Assembly using the Graph Theory. The readings produced by the sequence are split into windows and a network is formed, much more complex than that shown above. The algorithm next seeks to deduce a path passing through all nodes of the network only once (the dashed path is eliminated). The basic procedure to solve this problem already existed, before being used in genome assembly.

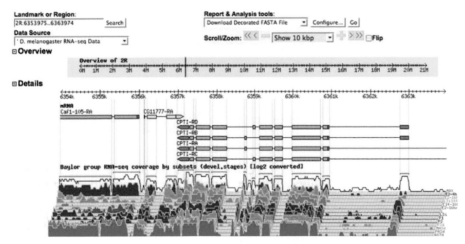

Figure 9.4 Expression profiles based on *RNAseq*, extracted from the database *FlyBase* (http://flybase.org), with the representation of the genome (upper scale), transcripts (*exons* represented by boxes and *introns* by lines), and the coverage of different regions by the anchored sequences, generated by next generation sequencing equipment.

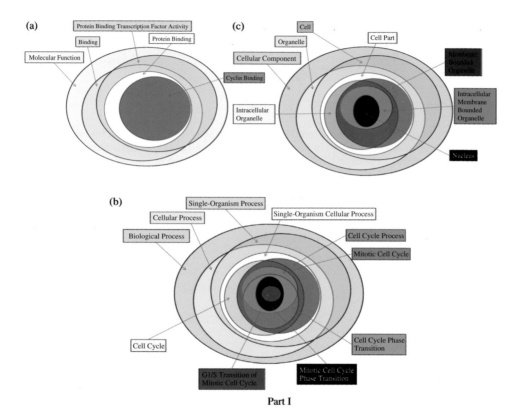

Part I

Figure 9.6 Gene Ontology. The hierarchy of *GO terms* related to (a) molecular function, (b) biological process, and (c) cellular component, associated with a protein that controls cellular proliferation, *cdk1*. Associated *GO terms* reveal that *cdk1* functions through cyclin binding, plays a role in G1/S transition of the mitotic cell cycle and occurs in the nucleus. It is easy to recover information associated to all children of a given *GO term*, for example, cell cycle (GO:0007049) (http://www.ebi.ac.uk/QuickGO/).

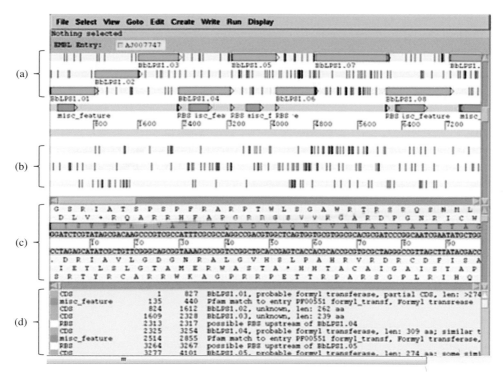

Figure 9.7 View of a window of software *Artemis*. This software allows the marking of regions in three reading frames of the two strands (a) and (b) in which nearby stop codons (vertical lines) do not occur, putative genetic regions (cyan) and visualization of the working amino acid sequence (c). Annotations based on *BLAST* can be edited in (d). Figure extracted from the *Artemis* website (http://www.sanger.ac.uk/resources/software/artemis).

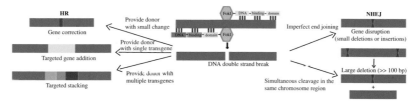

Figure 10.1 Modifications in specific DNA sequences are initiated with generation of a DNA double-stranded break (DSB). The diagram shows the possible results of a nuclease-mediated DSB. The presence of a sequence donor can stimulate the homologous recombination (HR) pathway, allowing donor sequence integration (on the left). DSBs also can be repaired by non-homologous end joining (NHEJ) (on the right), which can result in small insertions or deletions (indels) at the DNA break. These indels can interrupt the reading frame of the target gene, knocking it out. Two simultaneous DSBs made on the same chromosome can lead to a large deletion.

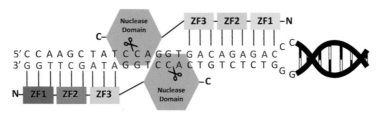

Figure 10.2 Scheme illustrating three ZFN modules, which recognize 9 bp upstream and 9 bp downstream of the region where the DSB is desired. Each ZFN module is linked to the catalytic domain of a nuclease that will catalyze the DSB.

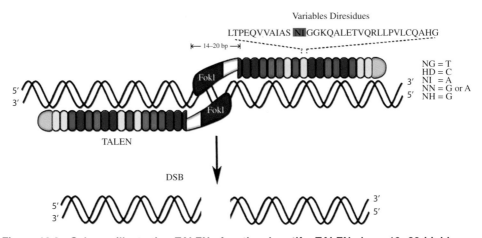

Figure 10.3 Scheme illustrating TALENs functional motifs. TALENs have 16–20 highly conserved repetitive peptidic modules, each of which is composed of 33 or 34 amino acids. The repeated modules differ at positions 12 and 13 (diresidues), which specifically recognize one of four nitrogenated bases.

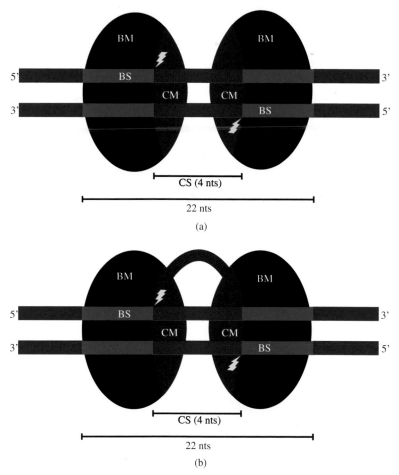

Figure 10.4 Meganucleases may have one (a) or two (b) copies of the motif LAGLIDADG, which is formed by the binding module (BM) and the cleavage module (CM). The BM recognizes the DNA binding site (BS). The CM acts at the DNA cleavage site (CS), creating two 3′-adhesive ends of four nucleotides each. Meganucleases with a single motif are active as homodimers, while versions with two motifs are active as monomers. (Source: Hafez and Hausner, 2012).

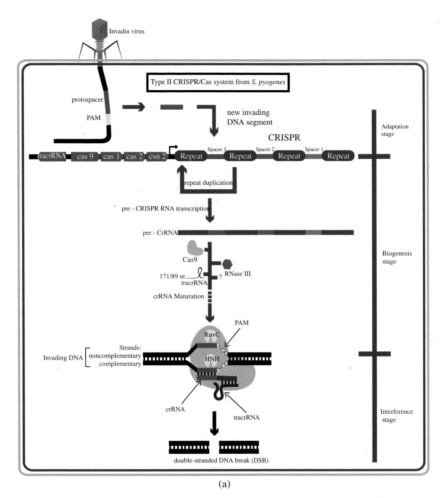

(a)

Figure 10.5 Type II CRISPR/Cas system pathway in *Streptococcus pyogenes* and targeted plant genome editing by type II sgRNA/Cas9 system. (a) Example of type II CRISPR/Cas system pathway in *S. pyogenes* in which an incorporation of a new invading DNA segment in the CRISPR locus occurs (adaptation stage); pre-CRISPR RNA transcription and crRNA maturation involving Cas9, RNase III, tracrRNA, and another possible unknown component (?) (biogenesis stage); and target DNA breaking by the tracrRNA–crRNA–Cas9 complex (interference stage). (b) Targeted plant genome editing by the type II sgRNA/Cas9 system involving coexpression of the cas9 host codon optimized protein bearing one nuclear localization signal (NLS) and gRNA expression by U6 polymerase III promoter.

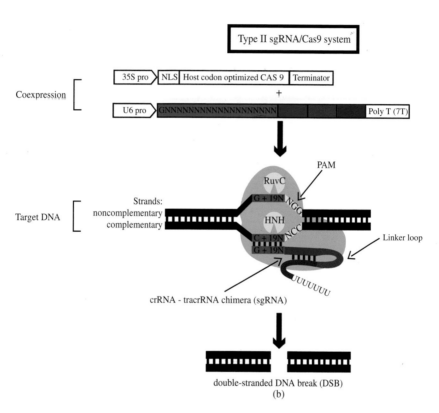

Figure 10.5 *(continued)*

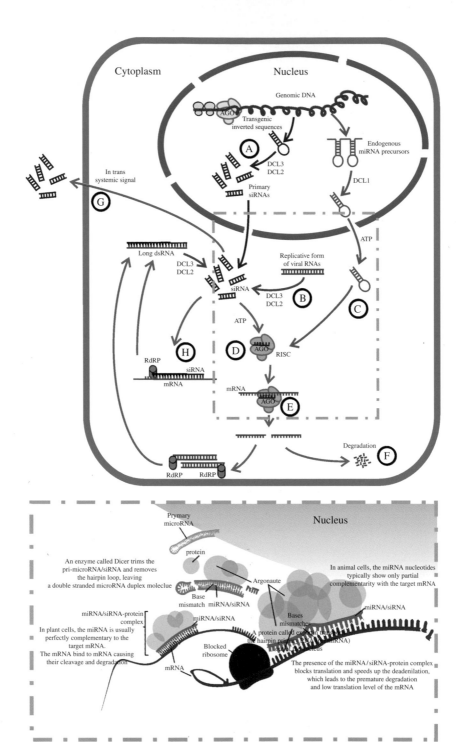

Figure 11.1 Gene silencing pathway. Dicer-like proteins processing transcripts containing inverted sequences (A), derived from viral RNA replication (B), and precursors of miRNA exported from the nucleus (C). Formation of siRNAs/RISC complex (D) directed to target RNA (E), which is subsequently, degraded (F); systemic silencing (G); and amplification by RdRP (H). (Source: Based on Souza *et al*., 2007; Aragão and Figueiredo, 2008).

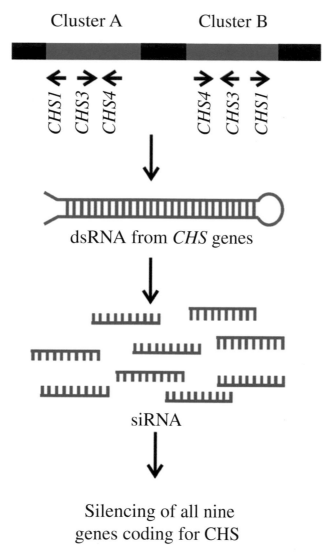

Cluster A   Cluster B

CHS1 CHS3 CHS4   CHS4 CHS3 CHS1

dsRNA from *CHS* genes

siRNA

Silencing of all nine
genes coding for CHS

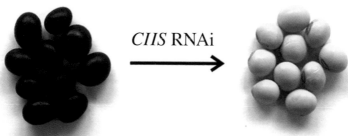

CHS RNAi

Figure 11.2   Silencing of *CHS* genes coding for chalcone synthase (a key enzyme in the biosynthesis of anthocyanins) in soybean. The presence of inverted repeat sequences of *CHS1*, *CHS3*, and *CHS4* leads to formation of RNA hairpin (dsRNA), which is processed to form siRNA, leading to silencing of all of the nine CHS genes. This manifests in the phenotypic characteristics of the seeds as colorless (yellow).

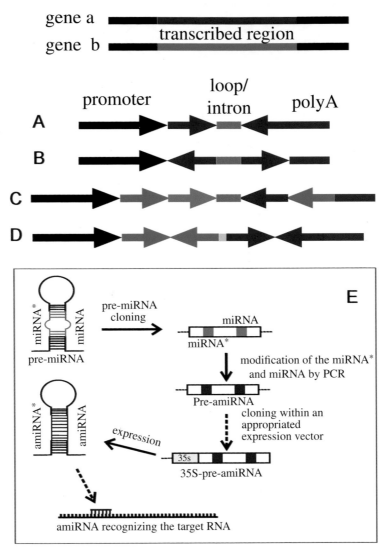

Figure 11.3 Vectors used in the stable transformation of plants are generally designed to produce hairpin structures (after transcription of RNA or dsRNA). Here, a transcribed sequence of a gene is amplified and placed under the control of a promoter in forward (sense) and reverse (antisense) directions spaced by an intron or a spacer region (loop) (A, B). Sequences of two or more genes can be used in the same expression cassete (C). It is also possible to join two expression cassettes harbouring gene fragments cloned in forward and reverse directions separated by a spacer (D). In (E), a vector designed to generate a modified miRNA by introducing a target gene sequence into a natural miRNA region (such as miR159 of *A. thaliana* and miR528 of *Oriza sativa*) is presented. The resulting vector expresses artificial miRNA (amiRNA) whose target may be the endogenous gene or that of an intracellular pathogen.

(3500 m above sea level) habitats have led to the evolution of endemic species with considerable resistance to drought (*S. pennellii, S.chilense*), salinity (*S. galapagense*) and cold (*S. habrochaites*) (Taylor, 1986).

This complex of species is a valuable source of quantitative trait loci (QTL) and of allelic variation for mendelian genes (Bai and Lindhout, 2007), which has already been studied for sugar metabolism in fruits (Chetelat *et al.*, 1995; Schaffer *et al.*, 2000; Fridman *et al.*, 2004) and organogenic capacity *in vitro* (Koornneef *et al.*, 1987; Faria and Illg, 1996; Arikita *et al.*; 2013). The observation of new phenotypes and identification of new alleles derived from wild species are made easier by the use of introgression lines (ILs), which are permanent mapping populations (Eshed and Zamir, 1994; Figure 6.1). Once identified, the specific effect of a given allele can be efficiently studied through the construction of

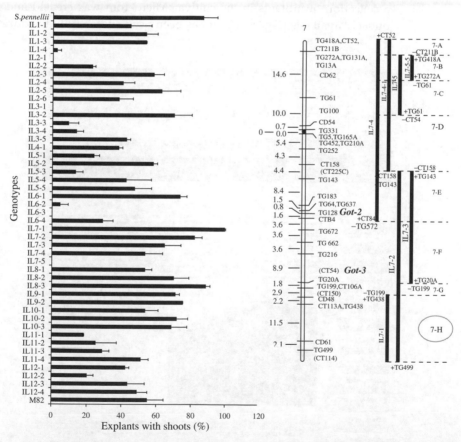

Figure 6.1   Demonstration of the use of introgression lines (ILs) containing small segments of the *S. pennellii* genome introgressed into the tomato cultivar M82. In this case, the lines were used to map a putative locus responsible for high *in vitro* regeneration capacity in *S. pennellii*. One of the genome segments containing genes for the said trait was found in bin H of chromosome 7, as it represents the intersection of two lines with high *in vitro* regeneration capacity, IL7-1 and IL7-2. (Source: Modified from Arikita *et al.*, 2013).

near-isogenic lines (NILs), which differ between them only in a small region around the QTL or mendelian locus (Paran and Zamir, 2003).

Taking into account that generation of NILs could demand several back-crossing generations, the use of a fast life cycle genotype, such as the tomato cultivar Micro-Tom (MT) is highly advantageous (Carvalho *et al.*, 2011). MT is a dwarf cultivar of determinate growth habit originally bred for ornamental purposes (Scott and Harbaugh 1989), which was subsequently adapted as a genetic model system (Meissner *et al.*, 1997; Marti *et al.*, 2006). The small size (8 cm if grown in 50–100 ml pots) puts it on a par with *Arabidopsis* in terms of practicality, with the added advantage that the tomato is an economically important crop, thus bringing together basic and applied research (Campos *et al.*, 2010).

At the Laboratory of Hormonal Control of Plant Development in the Biological Sciences Department of ESALQ/University of São Paulo, a collection of mutants in the cultivar Micro-Tom was created and is publicly available (Figure 6.2). It has been used successfully to address

Figure 6.2 Homepage of the Laboratory of Hormonal Control of Plant Development, where a large collection of tomato mutants in the cultivar Micro-Tom are curated http://www.esalq .usp.br/tomato.

various topics in plant physiology (Carvalho *et al.*, 2011), as well as for didactical purposes.

"Physionomics" as an Integrator of Various Omics for Functional Studies and Plant Breeding

As can be deduced from the topics developed earlier in this chapter, physiology is not a technique but rather a body of knowledge in itself, advanced by the appearance of new approaches, of which the "omics" is but the latest in a long series. Ultimately, because physiology aims to understand how plants function, the closest approach to "physionomics" would be "phenomics" (Furbank and Tester, 2009; Berger *et al.*, 2010; Tisné *et al.*, 2013). Phenomics is a series of high throughput techniques devised to automate and increase our capacity to identify and sort phenotypes; eventually explaining them mechanistically through their genes, transcripts, proteins, and metabolites (see Chapter 7).

Some examples of phenomic approaches are: (i) compound image capture systems coupled to software for automatic analysis of leaf area and rosette growth (Tisné *et al.*, 2013); (ii) infrared cameras for temperature gradient analysis, which indicates variation in energy dissipation and is of interest for drought response (Jones *et al.*, 2009; Munns *et al.*, 2010; Qiu *et al.*, 2009; Sirault *et al.*, 2009); (iii) fluorescence detection imaging systems to visualize differential response to stress factors, which can be used to screen large seedling populations (Badger *et al.*, 2009; Jansen *et al.*, 2009); (iv) methods for non-invasive underground structure visualization (Nagel *et al.*, 2012; Yazdanbakhsh and Fisahn, 2009); and (v) LIDAR (Light Detection and Ranging) technology to determine growth rate by laser measurement of tiny distance differentials (Hosoi and Omasa, 2009). It should be pointed out that the output of all of the aforementioned techniques is digital data, which can be relayed and stored remotely in servers for subsequent analysis, which is also usually carried out by custom software in an automated or semi-automated manner.

With scientific research being conducted on an ever larger scale and in an increasingly automated fashion, one expects that the role of the researcher will be more and more focused on experimental design, reflection, analysis, and theoretical insight. Thus, in the near future it can be envisaged that the large amount of data generated by the "omics" approaches will allow us to understand plants' responses to their environment, and, after sifting through the available physiological knowledge, lead to comprehensive breeding research programs that are directed towards sustainable increases in yield (Figure 6.3).

It should be noted that in the diagram shown in Figure 6.3, physiological knowledge is put forward as a pre-requisite for breeding. More to the point, some of the topics in plant breeding that could benefit from

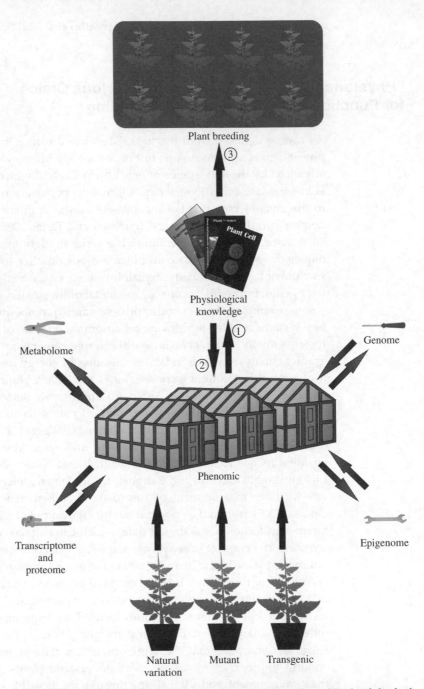

Figure 6.3 Physionomics as an integrator of various omics with physiological knowledge. The analysis of phenotypes (phenomics) can generate conceptual gains (1), and physiological knowledge can direct decision-making in breeding programs (3), besides improving our ability to analyze phenotypes (2). The omics tools must be complemented by focused studies *in planta*, represented here by mutants, natural genetic variation, and transgenic plants. Red arrows indicate classical genetics and blue arrows, reverse genetics.

plant physiology paradigms are: (i) generation of genotypes resistant to abiotic stress factors, particularly drought; (ii) generation of genotypes with resistance to biotic stresses, that is, pathogens; and (iii) generation of genotypes with higher yield, which could possibly be related to photosynthetic capacity.

In relation to the first item, useful theoretical physiological knowledge to guide the search for drought resistant plants was systematized in the 1970s, with the realization that different mechanisms are responsible for drought avoidance and drought tolerance. In the former, the plant "avoids" the stress, meaning that it does not equilibrate with its stressful environment, for example, it maintains high water potentials in its tissues even though the soil is dry. In the latter, the plant "tolerates" the stress, or, it equilibrates with the environment, thus the plant can cope with and continue to function even when its water potential drops considerably (Levitt, 1972).

Even though drought tolerance allows for the survival of xerophytes (arid tolerant plants) in hostile environments, the metabolic commitment required to do so compromises their usefulness in breeding programs. Thus, as appropriately pointed out by John Passioura, drought resistance in agriculture needs to be thought of in terms of yield and not just survival (Passioura, 1996). Whereas drought *tolerance* mechanisms relate to the generation of molecules protective of membrane integrity and function, as well as relief from oxidative stress (Thomashow, 1999), the mechanisms for drought *avoidance* are more complex and yet hold more promise for the production of genotypes with higher water-use efficiency (amount of dry matter produced per amount of water transpired during the growth cycle). Avoidance mechanisms tend to involve alterations in the hormonal control of development, which needs to fine-tune relative organ growth (Mingo *et al.*, 2004) and water loss (Dodd, 2003), so that the plants sustain water potentials allowing photosynthesis and growth under limited water supply (Sobeih *et al.*, 2003).

As for the generation of genotypes resistant to pathogens, this too is an area that can benefit greatly from recent advances in the understanding of hormone signaling mechanisms. The interaction between plant and pathogen during infection is controlled by signaling pathways. Necrotrophic and biotrophic pathogens are affected significantly by signaling pathways triggered by jasmonic acid (JA) and salicylic acid (SA), respectively (Glazebrook, 2005; McDowell and Dangl, 2000). Nevertheless, since JA and SA are antagonists (Glazebrook, 2005), it is clear that more in-depth knowledge of the signaling pathway of both hormones (Chini *et al.*, 2007; Thines *et al.*, 2007; Fu *et al.*, 2012) will be necessary to discover the specific stages in which JA-related signaling can be increased without decreasing SA signaling and vice versa. Otherwise, engineering the excess production of either hormone would lead to resistance to one type of pathogen (necrotrophic or biotrophic) while opening the door to the other (Robert-Seilaniantz *et al.*, 2007).

Finally, with respect to the search for more productive cultivars and their relationship to photosynthesis, an elegant experiment showed that wheat varieties released between 1900 and 1980, grown alongside each other in a single season, did not show any significant difference in total dry matter production at the end of their growth cycle (Gifford *et al.*, 1984). The fact that wheat cultivars released over an 80-year period did not differ in their dry matter production contradicts the notion that breeding allows the selection of genotypes with higher photosynthetic rates. Harvest index (kg grain/kg total dry mass), on the other hand, increased in direct proportion with yield in modern cultivars, particularly in those released from the 1950s onward, that is, starting with the Green Revolution.

This period was characterized by the introduction of semi-dwarf wheat and rice varieties with high yield, which besides the intrinsic yield increase brought about by a higher harvest index, allowed for the use of more fertilizers without causing lodging (Borlaug, 1983). As can be deduced by the Gifford *et al.* (1984) experiment, the aim of the Green Revolution was not to increase photosynthetic rate in crops, in line with the idea that an essential physiological process that arose more than 3.5 billion years ago is expected to have been optimized by nature itself. Ultimately, the Green Revolution came about by inadvertently manipulating hormonal control of plant development in order to change the relationship between source and sink and thus increase harvest index (Hedden, 2003). With the advances in our understanding of metabolism and hormone signaling pathways achieved by the use of *Arabidopsis*, it was possible to determine that the mutations selected in high-yield varieties reduce gibberellin biosynthesis in rice (Sasaki *et al.*, 2002), and reduce the sensitivity to that hormone in wheat (Peng *et al.*, 1999).

All of the examples described here show the potential that physiological knowledge, with the expected boost from the subsidiary "omics", has to direct breeding programs, making the search for superior genotypes less empirical. It should be taken into account, however, that the omics approaches can be overly dependent on model plants, as already discussed earlier. Such realization suggests reflection and caution when omics projects are put forward for plants less amenable to genetic studies, such as large, long-cycle ones or perennials, alogamous, polyploids, and species lacking biotechnological resources (e.g., efficient protocols for transformation, mutagenesis, and molecular markers). Moreover, the absence of mutants and tagged lines (with T-DNA, transposons or TILLING) in such species means that all of the genomics, transcriptomics, and proteomics efforts could not be backed by *in planta* information. In other words, whereas pinning large omics databases down to relevant genes or alleles is like finding a needle in a haystack, care must be taken so that investments in omics do not merely result in increasing the size of the haystack.

Acknowledgements

The authors thank Dr. Benjamin Rae (University of Oxford) and Jonata Freschi (University of São Paulo) for comments, suggestions, and editing.

References

Amasino, R. 2005. 1955. Kinetin arrives. The 50th anniversary of a new plant hormone. Plant Physiology, 138 (3): 1177–1184.

Arikita, F.N.; Azevedo, M.S.; Scotton, D.C.; *et al.* 2013. Natural genetic variation controlling the competence to form adventitious roots and shoots from the tomato wild relative *Solanum pennellii*. Plant Science, 199–200: 121–130.

Badger, M.R.; Fallahi, H.; Kaines, S.; Takahashi, S. 2009. Chlorophyll fluorescence screening of *Arabidopsis thaliana* for CO_2 sensitive photorespiration and photoinhibition mutants. Functional Plant Biology, 36 (11): 867–873.

Bai, Y.; Lindhout, P. 2007. Domestication and breeding of tomatoes: what have we gained and what can we gain in the future? Annals of Botany, 100 (5): 1085–1094.

Bassham, J.A.; Benson, A.A.; Kay, L.D.; *et al.* 1954. The path of carbon in photosynthesis. XXI. The cyclic regeneration of carbon dioxide acceptor. Journal of the American Chemical Society, 74 (7): 1760–1770.

Berger, B.; Parent, B.; Tester, M. 2010. High-throughput shoot imaging to study drought responses. Journal of Experimental Botany, 61 (13): 3519–3528.

Bevan, M.; Walsh, S. 2005. The Arabidopsis genome: a foundation for plant research. Genome Research, 15 (12): 1632–1642.

Borlaug, N.E. Contributions of conventional plant-breeding to food-production. 1983. Science, 219 (4585): 689–693.

Campos, M.L.; Almeida, M.; Rossi, M.L.; *et al.* 2009. Brassinosteroids interact negatively with jasmonates in the formation of anti-herbivory traits in tomato. Journal of Experimental Botany, 60 (15): 4347–4361.

Campos, M.L.; Carvalho, R.F.; Benedito, V.A.; Peres, L.E.P. 2010. Small and remarkable: The Micro-Tom model system as a tool to discover novel hormonal functions and interactions. Plant Signaling and Behavior, 5 (3): 267–270.

Carvalho, R.F.; Campos, M.L.; Pino, L.E.; *et al.* 2011. Convergence of developmental mutants into single tomato model system: Micro-Tom as an effective tool kit for plant development research. Plant Methods, 1 (7): 18.

Chang, C.; Kwok, S.F.; Bleecker, A.B.; Meyerowitz, E.M. 1993. Arabidopsis ethylene-response gene *ETR1*: similarity of product to two-component regulators. Science, 262 (5133): 539–544.

Chetelat, R.T.; Deverna, J.W.; Bennet, A.B. 1995. Introgression into tomato (*Lycopersicon esculentum*) of the *L. chmielewskii* sucrose accumulator gene (*sucr*) controlling fruit sugar composition. Theoretical and Applied Genetics, 91 (2): 327–333.

Chini, A.; Fonseca, S.; Fernández, G.; *et al.* 2007. The JAZ family of repressors is the missing link in jasmonate signalling. Nature, 448 (7154): 666–671.

Clouse, S.D.; Sasse, J.M. 1998. Brassinosteroids: essential regulators of plant growth and development. Annual Review of Plant Physiology and Plant Molecular Biology, 49: 427–451.

Dharmasiri, N.; Dharmasiri, S.; Estelle, M. 2005. The F-box protein TIR1 is an auxin receptor. Nature, 26 (7041): 441–445.

Dodd, I.C. Hormonal interactions and stomatal responses. 2003. Journal of Plant Growth Regulation, 22 (1): 32–46.

Emmanuel, E.; Levy, A.A. Tomato mutants as tools for functional genomics. 2002. Current Opinion in Plant Biology, 5 (2): 112–117.

Eshed, Y.; Zamir, D. 1994. A genomic library of *Lycopersicon pennellii* in *L. esculentum*: A tool for fine mapping of genes. Euphytica, 79 (3): 175–179.

Faria, R.T.; Illg, R.D. 1996. Inheritance of *in vitro* plant regeneration ability in the tomato. Brazilian Journal of Genetics, 19 (1): 113–116.

Fridman, E.; Carrari, F.; Liu, Y.S.; *et al.* 2004. Zooming in on a quantitative trait for tomato yield using interspecific introgressions. Science, 305 (5691): 1786–1789.

Fu, Z.Q.; Yan, S.; Saleh, A.; *et al.* 2012. NPR3 and NPR4 are receptors for the immune signal salicylic acid in plants. Nature, 486 (7402): 228–232.

Furbank, R.T.; Tester, M. 2011. Phenomics – technologies to relieve the phenotyping bottleneck. Trends in Plant Science, 16 (12): 635–644.

Gifford, R.M.; Thorne, J.H.; Hitz, W.D.; Giaquinta, R.T. 1984. Crop productivity and photoassimilate partitioning. Science, 225 (4664): 801–808.

Glazebrook, J. 2005. Contrasting mechanisms of defense against biotrophic and necrotrophic pathogens. Annual Review of Phytopathology, 43: 205–227.

Hareven, D.; Gutfinger, T.; Parnis, A.; *et al.*. 1996. The making of a compound leaf: genetic manipulation of leaf architecture in tomato. Cell, 84 (5): 735–744.

Harvey, R.B. 1929. Julius von Sachs. Plant Physiology, 4 (1): 154–157.

Hedden P. 2003. The genes of the Green Revolution. Trends in Genetics, 19 (1): 5–9.

Hosoi, F.; Omasa, K. 2009. Detecting seasonal change of broad-leaved woody canopy leaf area density profile using 3D portable LIDAR imaging. Functional Plant Biology, 36 (11): 998–1005.

Inoue, T.; Higuchi, M.; Hashimoto, Y.; *et al.* 2001. Identification of CRE1 as a cytokinin receptor from *Arabidopsis*. Nature, 409 (6823): 1060–1063.

Jansen, M.; Gilmer, F.; Biskup, B.; *et al.* 2009. Simultaneous phenotyping of leaf growth and chlorophyll fluorescence via GROWSCREEN FLUORO allows detection of stress tolerance in *Arabidopsis thaliana* and other rosette plants. Functional Plant Biology, 36 (11): 902–914.

Jones, H.G.; Serraj, R.; Loveys, B.R.; *et al.* 2009. Thermal infrared imaging of crop canopies for the remote diagnosis and quantification of plant responses to water stress in the field. Functional Plant Biology, 36 (11): 978–989.

Kepinski, S.; Leyser, O. 2005. The F-box protein TIR1 is an auxin receptor. Nature, 26 (7041): 446–451.

Koornneef, M.; Hanhart, C.J.; Martinelli, L. 1987. A genetic analysis of cell culture traits in tomato. Theoretical and Applied Genetics, 74 (5): 633–641.

Krysan, P.J.; Young, J.C.; Sussman, M.R. 1999. T-DNA as an insertional mutagen in Arabidopsis. The Plant Cell, 11 (12): 2283–2290.

Leonelli, S. 2007. *Arabidopsis*, the botanical *Drosophila*: from mouse cress to model organism. Endeavour, 31 (3): 34–39.

Levitt, J. Responses of Plants to Environmental Stresses. New York: Academic Press, 1972, 697 pp.

Li, J.; Nagpal, P.; Vitart, V.; *et al.* 1996. A role for brassinosteroids in light-dependent development of *Arabidopsis*. Science, 272 (5260): 398–401.

Libault, M.; Farmer, A., Brechenmacher, L.; *et al.* 2010. Complete transcriptome of the soybean root hair cell, a single-cell model, and its alteration in response to *Bradyrhizobium japonicum* infection. Plant Physiology, 152 (2): 541–552.

Lu, Y.; Last, R.L. Web-based Arabidopsis functional and structural genomics resources. 2008. The Arabidopsis Book, 6: e0118, doi: 10.1199/tab.0118.

Martí, E.; Gisbert, C.; Bishop, G.J.; *et al.* 2006. Genetic and physiological characterization of tomato cv. Micro-Tom. Journal of Experimental Botany, 57 (9): 2037–2047.

McDowell, J.M.; Dangl, J.L. 2000. Signal transduction in the plant immune response. Trends in Biochemical Sciences, 25 (2): 79–82.

Meissner, R.; Jacobson, Y.; Melamed, S.; *et al.* 1997. A new model system for tomato genetics. Plant Journal, 12 (6): 1465–1472.

Meyerowitz, E.M.; Pruitt, R.E. 1985. *Arabidopsis thaliana* and plant molecular genetics. Science, 229 (4719): 1214–1218.

Mingo, D.M.; Theobald, J.C.; Bacon, M.A.; *et al.* 2004. Biomass allocation in tomato (*Lycopersicon esculentum*) plants grown under partial root zone drying: enhancement of root growth. Functional Plant Biology, 31(10): 971–978.

Munns, R.; James, R.A.; Sirault, X.R.; *et al.* 2010. New phenotyping methods for screening wheat and barley for beneficial responses to water deficit. Journal of Experimental Botany, 61 (13): 3499–3507.

Myers, J. 1974. Conceptual developments in photosynthesis. Plant Physiology, 54 (4): 420–426.

Nagel, K. A.; Putz, A.; Gilmer, F.; *et al.* 2012. GROWSCREEN-Rhizo is a novel phenotyping robot enabling simultaneous measurements of root and shoot growth for plants grown in soil-filled rhizotrons. Functional Plant Biology, 39 (11): 891–904.

Panara, F.; Calderini, O.; Porceddu, A. 2012. *Medicago truncatula* functional genomics - An invaluable resource for studies on agriculture sustainability. In: Functional Genomics, Meroni, G. and Petrera, F. (eds). Rijeka, Croatia: InTech, doi: 10.5772/51016. Available from: http://www.intechopen.com/books/functional-genomics/medicago-truncatula-functional-genomics-an-invaluable-resource-for-studies-on-agriculture-sustainabi (accessed 1 December 2013).

Paran, I.; Zamir, D. 2003. Quantitative traits in plants: beyond the QTL. Trends in Genetics, 19 (6): 303–306.

Park, S.Y.; Fung, P.; Nishimura, N.; *et al.* 2009. Abscisic acid inhibits type 2C protein phosphatases via the PYR/PYL family of START proteins. Science, 324 (5930): 1068–1071.

Passioura, J.B. Drought and drought tolerance. 1996. Plant Growth Regulation, 20 (2): 79–83.

Peng, J.R.; Richards, D.E.; Hartley, N.M.; *et al.* 1999. 'Green revolution' genes encode mutant gibberellin response modulators. Nature, 400 (6741): 256–261.

Pnueli, L.; Carmel-Goren, L.; Hareven, D.; *et al.* 1998. The *SELF-PRUNING* gene of tomato regulates vegetative to reproductive switching of sympodial meristems and is the ortholog of *CEN* and *TFL1*. Development, 125 (11): 1979–1989.

Qiu, G.Y.; Omasa, K.; Sase, S. 2009. An infrared-based coefficient to screen plant environmental stress: concept, test and applications. Functional Plant Biology, 36 (11): 990–997.

Rick, C.M. Tomato mutants: freaks, anomalies, and breeder's resources. 1986. HortScience, 21: 917–918.

Robert-Seilaniantz, A.; Navarro, L.; Bari, R.; Jones, J.D.G. 2007. Pathological hormone imbalances. Current Opinion in Plant Biology, 10 (4): 372–379.

Saltveit, M.E.; Yang, S-F.; Kim, W.T. 1998. History of the discovery of ethylene as a plant growth substance. 1998. In: Kung, S.-D. and Yang, S.-F. (eds). Discoveries in Plant Biology, volume 01. Singapore: World Scientific Publishing Co. Pte. Ltd., pp. 47–70.

Sasaki, A.; Ashikari, M.; Ueguchi-Tanaka, M.; *et al.* 2002. Green revolution: A mutant gibberellin-synthesis gene in rice – New insight into the rice variant that helped to avert famine over thirty years ago. Nature, 416 (6882): 701–702.

Schaffer, A.A.; Levin, I.; Oguz, I.; *et al.* 2000. ADPglucose pyrophosphorylase activity and starch accumulation in immature tomato fruit: the effect of a *Lycopersicon hirsutum*-derived introgression encoding for the large subunit. Plant Science, 152 (2): 135–144.

Schmutz, J.; Cannon, S.B.; Schlueter, J.; *et al.* 2010. Genome sequence of the palaeopolyploid soybean. Nature, 463 (7278): 178–183.

Scott, J. and Harbaugh, B. 1989. Micro-Tom: A miniature dwarf tomato. Florida Agricultural Experiment Station Circular, 370: 1–6.

Sirault, X.R.R.; James, R.A.; Furbank, R.T. 2009. A new screening method for osmotic component of salinity tolerance in cereals using infrared thermography. Functional Plant Biology, 36 (10-11): 970–977.

Sobeih, W.Y.; Dodd, I.C.; Bacon, M.A.; *et al.* 2003. Long-distance signals regulating stomatal conductance and leaf growth in tomato (*Lycopersicon esculentum*) plants subjected to partial root-zone drying. Journal of Experimental Botany, 55 (407): 2353–2363.

Somerville, C.; Koornneef, M. 2002. A fortunate choice: the history of *Arabidopsis* as a model plant. Nature Reviews Genetics, 3 (11): 883–889.

Stevens, M.A.; Rick, C.M. 1986. Genetic and breeding. In: Atherton, J.G. and Rudich, J. (eds). The Tomato Crop: A Scientific Basis for Improvement. London: Chapman and Hall, pp. 35–109.

Takahashi, N. Discovery of gibberellin. 1998. In: Kung, S-D. and Yang, S.-F. (eds). Discoveries in Plant Biology, volume 01. Singapore: World Scientific Publishing Co. Pte. Ltd., pp. 17–32.

Taylor, I.B. 1986. Biosystematics of the tomato. In: Atherton, J.G. and Rudich, J. (eds). The Tomato Crop: A Scientific Basis for Improvement. London: Chapman and Hall, pp. 1–34.

The Arabidopsis Genome Initiative. 2000. Analysis of the genome sequence of the flowering plant *Arabidopsis thaliana*. Nature, 408 (6814): 796–815.

The Tomato Genome Consortium. 2012. The tomato genome sequence provides insights into fleshy fruit evolution. Nature, 485 (7400): 635–641.

Thines, B.; Katsir, L.; Melotto, M.; *et al.* 2007. JAZ repressor proteins are targets of the SCF(COI1) complex during jasmonate signalling. Nature, 448 (7154): 661–665.

Thoquet, P.; Ghérardi, M.; Journet, E.P.; *et al.* 2002. The molecular genetic linkage map of the model legume *Medicago truncatula*: an essential tool for comparative legume genomics and the isolation of agronomically important genes. BMC Plant Biology, 2: 1.

Tisné, S.; Serrand, Y.; Bach, L.; *et al.* 2013. Phenoscope: an automated large-scale phenotyping platform offering high spatial homogeneity. The Plant Journal, 74 (3): 534–544.

Thomashow, M.F. 1999. Plant cold acclimation: Freezing tolerance genes and regulatory mechanisms. Annual Review of Plant Physiology and Plant Molecular Biology, 50: 571–599.

Ueguchi-Tanaka, M.; Ashikari, M.; Nakajima, M.; *et al.* 2005. GIBBERELLIN INSENSITIVE DWARF1 encodes a soluble receptor for gibberellin. Nature, 437 (7059): 693–698.

Umehara, M.; Hanada, A.; Yoshida, S.; *et al.* 2008. Inhibition of shoot branching by new terpenoid plant hormones. Nature, 455 (7210): 195–200.

Warnock, S.J. Natural habitats of *Lycopersicon* species. 1991. HortScience, 26 (5): 466–471.

Yazdanbakhsh, N.; Fisahn, J. 2009. High throughput phenotyping of root growth dynamics, lateral root formation, root architecture and root hair development enabled by PlaRoM. Functional Plant Biology, 36 (11): 938–946.

Zsögön, A.; Lambais, M.R.; Benedito, V.A.; *et al.* 2008. Reduced arbuscular mycorrhizal colonization in tomato ethylene mutants. Scientia Agricola, 65 (3): 259–267.

7 Phenomics

Roberto Fritsche-Neto,[a] Aluízio Borém,[b]
and Joshua N. Cobb[c]

[a]Department of Genetics, University of São Paulo/ESALQ, Piracicaba, SP, Brazil
[b]Department of Crop Science, Federal University of Viçosa, Viçosa, MG, Brazil
[c]DuPont Pioneer, Johnston, IA, USA

Introduction

Plant breeders usually study the nature of genetic variation in segregating populations evaluated over different years and locations. Thus, with the use of appropriate experimental designs, researchers have obtained relevant information on the genetic diversity, heritability, environment influence, genetic value, and the best strategies to agronomically select superior crop varieties (Cobb *et al.*, 2013).

In the genomics era, emphasis has changed towards the evaluation of genetic diversity directly at the DNA level. In face of the decreasing cost of genotyping technologies and whole genome sequencing, there is strong motivation for the development of diversity panels and mapping populations in many species in order to understand the genetic control of complex traits and to discover useful genetics that can be easily deployed (McCouch *et al.*, 2012). This effort has been a catalyst for new ideas about how to manipulate genetic variation in order to characterize and create relevant germplasm (Chen *et al.*, 2011).

However, the availability of abundant genomic information has necessitated a fundamental shift in the research focus towards addressing the lack of high quality phenotypic information on relevant germplasm. Unlike genotyping, which is currently highly automated and uniform across different organisms, phenotyping is still a manual activity, tailored for each species, environment, and trait. It often incurs high labor costs, can be very sensitive to environmental fluctuations, and sometimes involves the integration of subjective assessments from different people. For these reasons phenotyping, at present, has become the main bottleneck of genetic analyses (Cobb *et al.*, 2013).

Omics in Plant Breeding, First Edition. Edited by Aluízio Borém and Roberto Fritsche-Neto.
© 2014 John Wiley & Sons, Inc. Published 2014 by John Wiley & Sons, Inc.

The phenotype of a plant is dynamic and often unpredictable because it represents a set of complex responses to endogenous and exogenous signals that are received cumulatively throughout the life history of an organism. Therefore, in order to understand the genetic variation that underlies complex traits, it is necessary to carry out phenotyping activities with both accuracy and precision. These activities must also be scalable and high-throughput in order to ensure that a large enough population size is screened over adequate representative target environments so as not to limit the genetic inferences that can be made (Weber and Broman, 2001). In this context, the development of phenotyping technology will likely parallel the developments that have been seen in genotyping technology, requiring increases in automation and collaborations between biologists, agronomists, engineers, and bioinformaticians. However, there will be a functional limit to the level of automation that can be imposed as the "phenome" cannot be discretized into single character units like the genome, but rather represents a continuous stream of information that becomes ever more informative as non-genetic sources of variation are continuously sampled over environments and years.

Phenomics or next-generation phenotyping is understood to be the pursuit of accurate (able to effectively measure traits without bias), precise (small variance between repeated measurements), relevant (of specific and meaningful value to stated objectives), and economical phenotyping efforts (Cobb *et al.*, 2013). Accordingly, phenomics involves a series of high-throughput techniques to increase and to automate the capacity and accuracy of phenotypic evaluation, so as to empower the discovery of the genes, transcripts, proteins, and metabolites (Tisné *et al.*, 2013) that interact with the environment to produce the biodiversity prevalent today and to drive the generation of new diversity that will serve to diminish the impact of climate change and other global challenges moving forward.

Examples of Large-scale Phenotyping

The integration of new technology into plant breeding programs is often a challenge for both public and private organizations, but the pace of technological innovations that have characterized the transition from the 20th to the 21st century has been so rapid that never before has it been so crucial to re-evaluate the breeding process to ensure success. This is particularly critical when considering phenotyping platforms because the economies of scale generally applied in a breeding program change the relative value of new approaches. While phenotyping has emerged as one of the major targets for innovation in the breeding pipeline, it is by no means the only port of entry. Successful breeding programs will

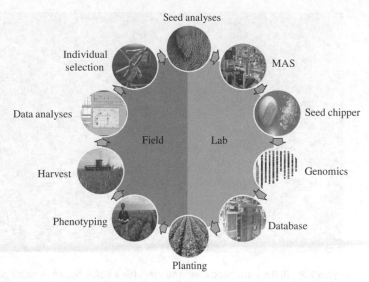

Figure 7.1   **Opportunities for the integration of new technologies into modern breeding programs.**

integrate modern technologies into several critical steps of the breeding process (Figure 7.1).

The development and use of tools to automate phenotyping, without sacrificing the power of prediction is one of the most critical points in the implementation of next-generation phenotyping (Cobb *et al.*, 2013). Over the last several years, improvements in automation, digital image analysis, and software development have created an opportunity for innovation of phenotypic analysis pipelines. This has spurned the creation of several phenotyping tools and automated systems now on the market and has contributed to a reduction in the demand for labor in laboratories, greenhouses, and in field experiments (Nagel *et al.*, 2012).

Some examples of phenomic approaches include: (i) the use of digital cameras to capture of zenithal images for automatic analysis of leaf area, canopy growth, and measurement of tissue, organs or individual traits (Tisné *et al.*, 2013); (ii) the use of infrared cameras to visualize stress response capacity and photosynthesis rates by recording temperature gradients, which indicate levels of energy dissipation (Munns *et al.*, 2010); (iii) the use of images produced by fluorescence detectors to visualize populations of plants with differential response to a stress conditions (Jansen *et al.*, 2009); (iv) the use of non-invasive methods to visualize roots and other below-ground systems (Nagel *et al.*, 2012); and (v) the use of the LIDAR (Light Detection and Ranging) technology to measure growth rate through a radar-like process that measures light reflectance rather than radio wave reflection (Hosoi and Omasa, 2009).

While not image based, one of the earliest phenotyping tools widely used to increase automation and throughput was the IRGA (Infrared

Figure 7.2   IRGA – Infrared Gas Analyzer. (See color figure in color plate section).

Gas Analyzer). Its importance is found in the easiness of estimating several physiological plant parameters simultaneously. Such traits include: net carbon assimilation rate, stomatal conductance, transpiration rate, chlorophyll *a* fluorescence, and maximum photochemical efficiency of photosystem II (Figure 7.2). This method permits a researcher to instantly obtain a view of the plant photosynthesis rate, by analyzing just one part of the leaf. Despite the fact that this data collection method sometimes does not reflect whole-plant photosynthetic capacity, it has proven a very useful and informative tool for research groups worldwide.

While the IRGA was a major step in high-throughput precise phenotypic data collection, it was not until technology permitted digital image analysis that the collection of quantitative data in automated or semi-automated ways became a real possibility. Digital image analysis has allowed many aspects of plant development, function, and health to be measured and tracked over time. This is particularly powerful when considering the fact that the same organism can be phenotyped several times over the course of its lifetime, thus empowering in-depth investigations into developmental phenotypes and their reaction to environmental stress.

Interestingly, the collection of image data, while a challenge in and of itself, is not the major bottleneck in employing this phenomic strategy. The approach is really driven by the increased computational and storage capacity of modern computers and the development of novel software that permits the processing and extraction of quantitative phenotypic data from digital images. A good example of this would be the notoriously difficult analysis of plant roots. Using images obtained by

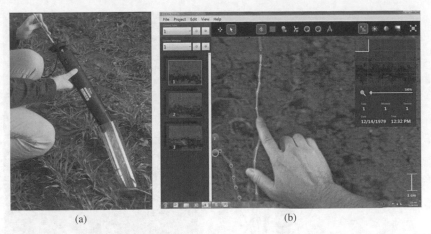

(a)   (b)

Figure 7.3   Scanner for collection of root images (a) and *output* of RootSnap software (b). (See color figure in color plate section).

scanners, one software package, known as RootSnap, simultaneously estimates root length, surface area, and root volume (Figure 7.3).

One of the major advantages of phenotyping by digital image analysis is the ability to quantify multiple phenotypic measurements from a single image or a group of images taken in a time series. An example is the methodology described well by Zia *et al*. (2012). These workers developed a tool for identification of corn genotypes tolerant to water stress. To this aim, image collection is initially carried out using infrared thermal sensors, such as the equipment at Figure 7.4.

From the images collected, data are obtained and processed for identification of the most promising genotypes (Figure 7.5).

According to the same workers, higher leaf temperature is associated with higher stomatal conductance, productivity, and plant height under

Figure 7.4   Image collection of an experiment. (Source: www.lemnatec.de. Reproduced with permission of LemnaTec GmbH, Aachen, Germany). (See color figure in color plate section).

Figure 7.5   Infrared thermal image of a genotype under stress condition, show-ing temperature distribution along measured plant. (Source: www.lemnatec.de. Reproduced with permission of LemnaTec GmbH, Aachen, Germany). (See color figure in color plate section).

drought conditions. Therefore, this tool may be particularly useful to breeding programs targeting drought tolerance. It is suggested that remotely flown drones, equipped with a digital camera may be used for larger scale data collection.

The advancements in large-scale phenotyping are not limited just to field grown applications, but are also of particularly meaningful value in greenhouses and growth chambers. This is exemplified by the phenotyping platforms offered by some companies, such as Lem-naTec (www.lemnatec.de; accessed 2 July 2013). Their system offers high-throughput, quantitative, and non-destructive image analysis of plants in multiple wavelengths, including infrared. This can be repeated at several intervals along the plant's lifecycle (Figure 7.6), allowing for the modeling of plant development dynamically over time.

The result is a series of image data from which many aspects of plant development can be inferred. This allows a quantitative representation, both for modeling purposes, and for identification and quantification of physiological and genetic parameters controlling plant development. Furthermore, all evaluations are carried out under high levels of envi-ronmental uniformity and allow the study of the effects of other factors on the plant, such as pulverization of pesticides, herbicides, or pathogen inoculation.

(a)

(b)

Figure 7.6   Scanalyzer³ᴰ Discovery Platform. Plants are taken individually by conveyor belt to the image collection cabin (a), where a digital image is collected from multiple angles (b). (Source: www.lemnatec.de. Reproduced with permission of LemnaTec GmbH, Aachen, Germany). (See color figure in color plate section).

For more field based phenotyping strategies, small plot sized automated harvesters have been used to enable breeders to carry out large numbers of trials with thousands of entries, with relative ease. In some cases, this has enabled a dramatic increase in the capacity of these breeding programs. Many harvesters and planters available on the market today come with automated processes that reduce or eliminate the possibility of human error. These include, but are not limited to: bar coded seed packages that identify each hill; GPS guided planting; handheld computers that associate bag number with geo-referenced plot locations; as well as weighing systems and grain moisture meters. Data from these instruments can often be monitored in real-time using computer equipment next to the driver in the cab of the harvester itself. Some combines even come with a system to send labeled and bagged samples directly into a hopper for unloading later, saving valuable time during harvest. Many combines also have the option to analyze the material content by means of near-infrared spectrophotometry (NIRS) and to take phenotypic measurements of plots prior to destructive harvesting.

One major advantage of these systems is that they allow remote data storage and transmission on internet connected servers, empowering subsequent automated analysis performed by scientists at a nearby laboratory or on the other side of the globe.

Important Aspects for Phenomics Implementation

In order for the proper integration of phenomics into breeding or genetics programs, the approach needs to be efficient and demonstrate the ability to increase genetic gain or genetic signal. In order to achieve efficiency, there are many questions that must be asked and decisions that must be made regarding accuracy, precision, automation, relevance, and cost of the phenotyping process. These questions and their answers may range from cultivation techniques, to choice of experimental designs, to management of procedures designed to collect and analyze data. It is important to consider the balance between automation, accuracy, and precision, with adaptability and relevance to actual breeding populations (Figure 7.7).

Sampling versus Controlling Environmental Variation

It is important to emphasize that for different purposes various phenotyping procedures are necessary. For example, under field conditions, it is generally convenient to group phenotypes into classes to facilitate real time data collection, at reasonable cost. This has been a long-time practice among plant breeders, who typically work with large populations. For this, several standardized and easy-to-apply scales have been developed

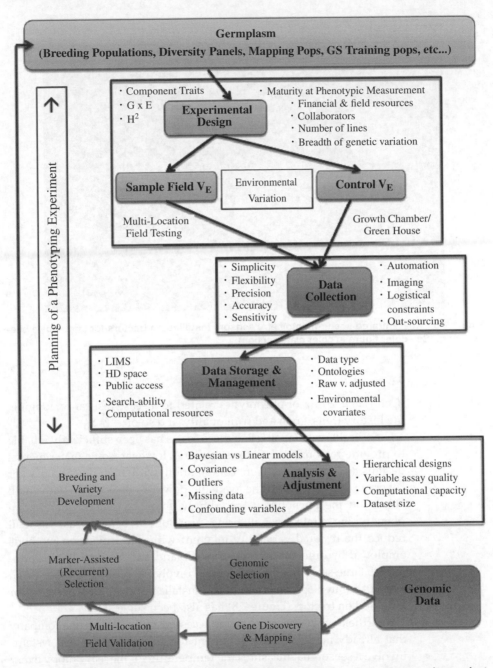

**Figure 7.7   Main considerations for planning and executing a large-scale phenotyping project.
(Source: Adapted from Cobb _et al._, 2013).**

Figure 7.8   Near-infrared spectrophotometry sensors installed on tractors for large-scale phenotyping. (See color figure in color plate section).

(Cobb *et al.*, 2013) for phenotypes related to disease resistance, lodging, life history traits, grain and fruit quality, and so on.

Historically, evaluation using these scales has been sufficiently reliable to provide reasonable data for selection. However, some experimental designs (Yu *et al.*, 2008), in combination with high-density marker coverage, have increased the available power to detect QTL with small effects on the phenotype. Accordingly, more consistent quantitative phenotyping strategies will bring pronounced benefits, as this will reduce the residual variability inherent with data collection methods employed by different investigators (Poland and Nelson, 2010). Automated large-scale field phenotyping involving remote sensors (such as near-infrared spectrophotometry) installed above the experiment or mounted on tractors (Figure 7.8) has also been suggested.

Smartfield (www.smartfield.com; accessed 2 July 2013) is one company that already markets such equipment for this purpose. Their systems involve a sensor that measures the temperature of the leaf canopy every 15 minutes and transmits this information to the base station. As soon as this information is loaded from the cellular tower to the server, the researcher can access the results via the internet (Figure 7.9). Such a system can allow for the identification of soybean diseases up to 5 days before the visual symptoms appear, or to quantify stress levels in real time accumulated by the plants, and aid in yield estimation.

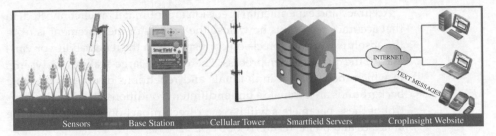

Figure 7.9   Scheme of a SmartField phenotyping platform. (Source: www.smartfield.com).

One important point to remember when considering these technologies is that in order to perform a correct characterization of the plants under evaluation it is necessary to calibrate the image processing and interpretation systems. This may involve the use of small populations under different stress levels, cultivation patterns, and/or locations. The pattern of plant response to that stress is then used to evaluate the relative performance of different populations at different plant development stages. However, this requires a significant initial investment to develop an accurate model (Cobb *et al.*, 2013).

The use of managed or controlled environments may be of some real value in reducing the phenotyping load by allowing a researcher to survey broad ranges of genetic variation with few other influences, and developing an approach to field phenotyping with targeted germplasm. By controlling irrigation time, amount of fertilizers, and plant health, individuals can be exposed, for example, to contrasting levels of a specific stress factor during their development (Cobb *et al.*, 2013). These strategies work particularly well when the heritability is high. In this context, breeders need to consider several aspects of each platform and whether they support their experimental objectives (Table 7.1).

Table 7.1   Factors to be considered in choice of phenotyping platform: controlled conditions versus field conditions.

Controlled conditions	Field conditions
Minimizes environmental variation and increases heritability	Maximizes relevance for breeders and growers
Increases precision of critical estimates	Characterizes environmental variation spectrum
Maximizes information with minimum repetitions	Evaluates in different times and places
Reduces costs through automation and standardization	Estimates G × E interaction
Develops hypotheses to be tested in field lines	Refines hypothesis and develops new evaluation protocols

Source: Adapted from Cobb *et al.* (2013).

Accuracy and experimental precision are intimately interconnected, in that accuracy represents how close the process or measurement is from the absolute truth and precision corresponds to the repeatability or variance of the quantification process. Therefore, in large-scale phenotyping, in which thousands of individuals and repetitions of different genetic backgrounds are evaluated under different conditions, the effective control of these parameters will have a direct impact on the quality of the analyses and on the results (Cobb *et al.*, 2013).

Unfortunately, in large-scale phenotyping strategies there will always be a trade-off between accuracy/precision and throughput/cost. As scale and standardization increase, it is necessary to think carefully about where losses in accuracy and precision can reasonably be incurred to the benefit of cost and throughput without biasing your results. It is not always easy to establish this appropriate balance, but with innovative designs, phenotyping tools can be developed that meet the purposes of breeders and are economically feasible.

Storage and Data Management

To accompany the exponential increase in data volume incumbent upon these high throughput strategies, there is also the need for a parallel increase in electronic storage capacity, processing, and analysis. While plant biology is not the first discipline to endure this challenge, simply storing the data with no mechanism to extract biological and agronomic relevance defeats the purpose of the endeavor itself. The research community as a whole will be required to come together around commonly agreed upon portals for data storage and analysis in order to leverage each other's scientific contributions. Advances in bioinformatics that have taken place over the last few decades will be helpful but more creative thinking will be required in order to accommodate the complex and inter-related nature of phenotypic information.

Organizing this information into a phenome (analogous to assembling a genome) will be a much bigger challenge due to the continuous, multifaceted, and subjective nature of phenotypic data in general (National Science Foundation, 2011). Consequently, databases designed to house phenotypic data can (and should) apply the lessons learned through the genomic era as DNA sequences became evermore available (Stein, 2010), but it will still face key bottlenecks, which genomics has never had to deal with.

In order to meet the needs of this growing mass of phenotypic information, sophisticated computational methods are necessary to store, analyze, and interpret correctly the complex and highly structured information that phenotype data represents. If not done carefully, critical information can be lost when data descriptions are summarized in only a few terms. In line with this, there are some specialized database projects specializing in plant phenomics. One of them is PHENOPSIS

DB (Juliette *et al.*, 2011), which hosts information on growth responses of *Arabidopsis thaliana* to different environmental conditions. This database has phenotypic information extracted from images and automated measurements collected in special growth chambers.

Another good example would be the Phenoscape project (http://kb .phenoscape.org), which has developed algorithms for semantic searches able to associate biological data by relationships between terms and hierarchical structures imposed by encoded descriptions. In this system data are characterized as sentences, where there is a subject (e.g., "a floret"), a predicate (e.g., has the color), and an object (e.g., white). Thus, the capturing of phenotypic metadata using this strategy adds the required dimensionality for the discovery of biological meaning, using linguistics and intuitive tool sets (Cobb *et al.*, 2013).

This underscores the point that the use of controlled vocabularies is an important characteristic to be shared among research communities, and likewise needs to be shared between these databases as well. The use of standardized expressions in such a way is called an "ontology." These ontologies can be used as tags to qualify and describe features related to biological data such as traits, genes, environments or taxonomy. For example, the hierarchical terms "growth and development ≥ aerial part development ≥ inflorescence development ≥ flowering time ≥ days to flowering" can be formally used to describe what is colloquially referred to as "flowering time." In this context, the most used ontologies include the plant ontology (PO), plant trait ontology (TO), plant environment ontology (EO) and phenotypic quality (PATO) (Cobb *et al.*, 2013).

Statistical Analysis of Data

With increased automation and scale and decreased labor costs comes the need for more specialized experimental designs that preserve and leverage predictive power. This can be observed in particular when the analyses involve many images originating from different individual groups, but need to be studied using similar algorithms and predefined commands in order to create a composite dataset.

Except for the challenge of storing terabytes worth of data in a sustainable way, the quality of images, in general, does not constitute a problem during large-scale phenotyping. However, if any individual deviates from patterns assumed by measurement algorithms, unpredictable and misleading values may appear. Thus, even with exhaustively tested automated analysis algorithms, it is necessary that a researcher regularly verifies and manually validates any system designed to collect such data (Clark *et al.*, 2012).

The quality of phenotypic data sets is often as valuable, or even more valuable, as the inferences made from them. This is because they are often used and re-used in different contexts to ask different questions.

Therefore, caution must be taken in making these inferences, so as to correctly characterize relationships. This is particularly meaningful when considering the genotype–phenotype relationship. Correct estimates of genetic gains by selection, for example, strongly depend on precise heritability and covariance estimates between phenotypes (Dickerson, 1955). Since one of these parameters is not directly observable, it needs to be estimated from phenotypic data using a variety of statistical models. As such, linear models have long been the basis of experiments in quantitative genetics (Lynch and Walsh 1997). These are popular because of their simplicity and the wide variety of easy-to-use software available to implement them. Nevertheless, these methods do have certain limitations.

Fundamentally, these models estimate parameters by identifying and returning a maximum likelihood value while ignoring the inherent uncertainty underpinning the values themselves. The parameters are then tested for statistical significance, based on specific significance levels (typically 5%), and then inferences are made based on whether or not the data meet these arbitrarily defined thresholds. More specifically, these estimates are good only when measurements are extensively repeated and follow a normal distribution. Thus, a great variety of procedures for data normalization and deletion of atypical figures is necessary to meet the considerations assumed by the model. Unfortunately, these methods frequently present statistical limitations as they specify arbitrary levels for data exclusion.

In contrast, the use of Bayesian strategies for statistical inferences is fundamentally different and compensates for the limitations of maximum likelihood models. Instead of approaching a local maximum that "best" estimates the parameter space, the purpose of Bayesian inference is to embrace the inherent variation in parameter estimates, and take into account prior information that may be useful in determining parameter values. This perspective on inference is much more aligned with biological reality and is preferred when working with hierarchically nested phenotypic data highly contextualized by both genetic and environmental variation (Cobb et al., 2013).

The disadvantage of Bayesian inference is its computational complexity, but with the increase of computational processing capacity, it is now possible to apply this strategy even when the data set is large and multifaceted. Additionally, there are now several software packages that allow the construction of Bayesian models in a relatively simple framework that are accessible to scientists with little or no experience in programing (Lunn et al., 2009).

Finally, in cases involving atypical designs, Bayesian models can also overcome some limitations of maximum likelihood testing (Greenberg et al., 2010). This is because they are easy to expand to nested hierarchical models, they are more robust to violations of normality and balanced designs, they can easily incorporate test quality metrics to weight the variables in analysis, and they can integrate across multiple phenotypes by effectively employing multivariate models. Finally, while nothing can substitute for repeated measures and good experimental design, Bayesian methods can borrow information from across the dataset to reduce the inherent uncertainties in model parameters and as such can reduce the need for extensive biological repetition, which can allow for the testing of a greater number of entries than could otherwise be evaluated (Greenberg *et al.*, 2010).

Main Breeding Applications

When combined with careful genotyping of thoughtful germplasm and the creation of targeted populations, next-generation phenotyping can both facilitate the discovery of genes as well as improve gain from selection in a breeding context. Its application ranges from germplasm characterization, to genomic selection (GS), to QTL identification.

Germplasm Characterization

Germplasm characterization is of great importance for plant breeding and genetics. Activities under this umbrella range from understanding how natural and breeding populations are structured in time and space to the creation of specific populations for gene discovery and allele mining.

The characterization of genetic diversity used to fall squarely in the domain of phenotypic evaluation. Collections of diverse lines are often evaluated for morphological traits and agronomic performance, particularly as they are multiplied from gene banks collections, for rejuvenation. Nevertheless, recent advancements in molecular biology have opened new frontiers for the disciplines of species conservation and population biology. With the use of molecular markers, the detection of the existing variability directly in the DNA has been made possible.

At present, the large scale genotyping of the world's germplasm collections is not a reality, but as funding and intellectual resources are applied in this sphere over the next decade, it will become possible to genotype large numbers of accessions in the gene banks worldwide, which will empower the mining of these tremendous resources as the genetic variants that are discovered become associated with phenotypes of interest. This will effectively unlock this genetic resource and enable the exploration of the genetic architecture of complex biotic and abiotic stresses. Careful thought about phenomics and how to apply it to the characterization of our crop diversity can increase the probability of finding otherwise rare alleles that can revolutionize global breeding efforts.

Genomic Selection (GS)

Throughout the years plant breeders have prioritized simplicity, speed, and flexibility over precision, accuracy, and automation in plant evaluations. This is because, historically, the advantages of the latter did not improve genetic gains as quickly. As genomic selection is applied to breeding programs, this is beginning to change.

While the efficiency in collecting large-scale genomic information increases and the costs decrease compared with phenotyping, plant breeders are starting to realize that genomic information can be a useful and economical approach to predicting phenotypic breeding value (Cabrera-Bosquet *et al.*, 2012). Thus, if the accuracy of genomic predictions is sufficient to justify the time required to do it and reduce the costs compared with evaluating segregating populations by the traditional method, and if GS demonstrates significant increases in genetic gain in a year by reducing cycle time, then plant breeders will eventually fully incorporate this strategy into their programs.

It is worth emphasizing that the objective of genomic selection is to use SNP variation in a population without considering the significance of marker effects to predict how suites of different alleles collectively contribute to the phenotype. Consequently, the objective is to predict the performance of a genotype under conditions representative of the target population of environments. For this, the following considerations should be observed when training a prediction model: (i) phenotypic estimates for individual performance should be done under field conditions; (ii) the number of locations should adequately sample the extent of environmental variation over which a breeder wishes to predict; and (iii) the cost of accurately phenotyping the training population across as many locations as necessary should be less than or comparable to the cost of collecting direct phenotypic observations on the entire breeding population. It is critical to keep in mind that these cost savings may come in the form of decreased cycle time but at the expense of selection accuracy (Cobb *et al.*, 2013).

For GS applications three populations need to be defined: a training population, a validation population, and a selection population. They can: (i) be physically different (three different populations); (ii) practice two functions at the same time (only one population is used for estimation and validation); or (iii) practice three functions at the same time (only one population is used for estimation, validation, and selection). In general, strategies (i) and (ii) are most frequently adopted (Figure 7.10; Resende *et al.*, 2010).

In GS models, precision phenotyping is important in the evaluation of the training population, since it provides the basis for prediction of phenotypic performance of the breeding population. This population contemplates a great number of markers evaluated at a moderate number of individuals (1000–2000, depending on the desired accuracy),

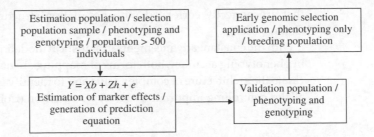

Figure 7.10 Scheme of strategy (ii) for genomic selection application in a breeding program. (Source: Resende *et al.*, 2010).

which must have their phenotypes evaluated according to traits of interest. Genomic prediction equations are then obtained for each considered trait, and used to estimate a GEBV (genomic estimated breeding value; Resende *et al.*, 2010).

Additionally, it is important to emphasize that the ability of genomic selection in predicting phenotypes with any accuracy depends on the use of large and representative training populations. This is especially true when the traits are of low heritability and of complex inheritance (Guo *et al.*, 2012; Resende *et al.*, 2012). This is due to the fact that the genotype by environment (G × E) interaction has a real impact on field performance, and GS is highly dependent on the development of models from individuals grown for a limited number of environments. Further studies are necessary to increase accuracy of GS models over time.

QTL Identification

In contrast to genomic selection, phenotyping of diversity panels for Genome Wide Association Studies (GWAS) or of bi-parental mapping populations for QTL analysis is designed to dissect the genetic architecture of complex traits and to understand the underlying genetic variation that drives the inheritance of important phenotypes (Cobb *et al.*, 2013). Both forms of mapping analysis are essential in the determination of genomic regions controlling complex trait variation. Cloning genes and discovering the functional polymorphisms that underlie QTL, however, is still an onerous and laborious activity, even when a QTL explains a substantial portion of the phenotypic variance (Saito *et al.*, 2010).

Biparental populations have the limitation of testing only alleles polymorphic between the two parents. Even so they offer great dissection power for QTL because they are not confounded by population structure and maintain allele frequencies at or near 50%. Conversely, association mapping studies can provide high QTL resolution for the same number of lines and evaluate a wider allele frequency spectrum, but manifest an inability to investigate rare alleles or dissect phenotypes that are perfectly correlated with population structure (Manolio and Collins, 2009).

With the constant advancements in statistical modeling and an increase in phenotyping accuracy, both forms of mapping are needed to fully elucidate the architecture of complex traits and to identify genes and specific alleles controlling important phenotypic variation (Cobb *et al.*, 2013).

Final Considerations

As phenomics is integrated with other omics technologies, significant strides can be made in understanding the "black box" of phenotypic variation and answer many questions on the origin of phenotypic diversity. Such questions include the following. To what extent does environmental variation determine the phenotypic effect of a given genetic variant? What is the role of epistasis and epigenetics in the determination of phenotypic variation? How can genetic information best be used to predict phenotypic performance and select germplasm that will form the basis of the world's food supply? For these aims, great effort and collaborative work among scientists in several disciplines will be necessary. Agronomists, biologists, physicists, engineers, and computer scientists will have to work together like they never have before. Moreover, it is always important to bear in mind that simply collecting massive amounts of data is not enough, but rather using that information as a tool to deliver meaningful biological and agronomic interpretations.

References

Cabrera-Bosquet, L.; Crossa, J.; von Zitzewitz, J; *et al.* 2012. High-throughput phenotyping and genomic selection: the frontiers of crop breeding converge. Journal of Integrative Plant Biology, 54: 312–320.

Chen, H.; He, H.; Zou, Y.; *et al.* 2011. Development and application of a set of breeder-friendly SNP markers for genetic analyses and molecular breeding of Rice (*Oryza sativa* L.). Theoretical and Applied Genetics, doi:10.1007/s00122-011-1633-5.

Clark, R.T.; Famoso, A.N.; Zhao, K.; *et al.* (2012) High throughput two dimensional root system phenotyping platform facilitates genetic analysis of root growth and development. Plant Cell & Environment, doi:10.1111/j.1365-3040.2012.02587.x.

Cobb, J.N.; DeClerck, G.; Greenberg, A.; *et al.* 2013. Next-generation phenotyping: requirements and strategies for enhancing our understanding of genotype–phenotype relationships and its relevance to crop improvement. Theoretical and Applied Genetics, 126: 867–887; doi: 10.1007/s00122-013-2066-0.

Dickerson, G. (1955) Genetic slippage in response to selection for multiple objectives. Cold Spring Harbor Symposium on Quantitative Biology, 20: 213–224.

Greenberg, A.J.; Hackett, S.R.; Harshman, L.G.; Clark, A.G. 2010. A hierarchical Bayesian model for a novel sparse partial Diallel crossing design. Genetics, 185: 361–373.

Guo, Z.; Tucker D.M.; Lu, J.; *et al.* 2012. Evaluation of genome-wide selection efficiency in maize nested association mapping populations. Theoretical and Applied Genetics 124: 261–275.

Hosoi, F.; Omasa, K. 2009. Detecting seasonal change of broad-leaved woody canopy leaf area density profile using 3D portable LIDAR imaging. Functional Plant Biology, 36 (11): 998–1005.

Jansen, M.; Gilmer, F.; Biskup, B.; *et al.* 2009. Simultaneous phenotyping of leaf growth and chlorophyll fluorescence via GROWSCREEN FLUORO allows detection of stress tolerance in *Arabidopsis thaliana* and other rosette plants. Functional Plant Biology, 36: 902–914.

Juliette, F.; Myriam, D.; Vincent, N.; *et al.* 2011. PHENOPSIS DB: an information system for *Arabidopsis thaliana* phenotypic data in an environmental context. BMC Plant Biology, 11: 77.

Lunn D.; Spiegelhalter D.; Thomas A.; Best, N. 2009. The BUGS project: evolution, critique and future directions. Statistics in Medicine, 28: 3049–3067.

Lynch M.; Walsh, B. (1997) Genetics and Analysis of Quantitative Traits. Sunderland, MT: Sinauer Associates Press, 980 pp.

Manolio, T.A.; Collins, F.S. 2009. The hapmap and genome-wide association studies in diagnosis and therapy. Annual Review of Medicine, (60): 443–456.

McCouch, S.R.; McNally, K.L.; Wang W.; *et al.* 2012. Genomics of gene banks: a case study in rice. American Journal of Botany, 99: 407–423.

Munns, R.; James, R.A.; Sirault, X.R.; *et al.* 2010. New phenotyping methods for screening wheat and barley for beneficial responses to water deficit. Journal of Experimental Botany, 61 (13): 3499–3507.

Nagel, K.A.; Putz, A.; Gilmer, F.; *et al.* 2012. GROWSCREENRhizo is a novel phenotyping robot enabling simultaneous measurements of root and shoot growth for plants grown in soil-filled rhizotrons. Functional Plant Biology, 39 (11): 891–904.

National Science Foundation. 2011. Phenomics: genotype to phenotype. St Loius: USDA and NSF Press, 42 pp.

Poland J.; Nelson, R. 2010. In the eye of the beholder: the effect of rater variability and different rating scales on QTL mapping. Phytopathology. doi:10.1094/PHYTO-03-10-0087.

Resende.; M.D.V.; Resende, M.F.R. Jr.; Aguiar .; *et al.* 2010. Computação da seleção genômica ampla (GWS). Colombo: Embrapa Florestas, 79 pp.

Resende, M. Jr.,; Munoz, P.; Resende, M.D.V.; *et al.* 2012. Accuracy of genomic selection methods in a standard data set of loblolly pine (*Pinus taeda* L.). Genetics, 190: 1503–1510.

Saito, K.; Hayano-Saito, Y.; Kuroki, M.; Sato, Y. 2010. Map-based cloning of the rice cold tolerance gene Ctb1. Plant Science, 179: 97–102.

Stein, L.D. 2010. The case for cloud computing in genome informatics. Genome Biology, 11: 207.

Tisné, S.; Serrand, Y.; Bach, L.; *et al.* 2013. Phenoscope: an automated large-scale phenotyping platform offering high spatial homogeneity. The Plant Journal, 74 (3): 534–544.

Weber, J.L.; Broman, K.W. 2001. Genotyping for human whole genome scans: past, present, and future. Advances in Genetics, 42: 77–96.

Yu, J.; Holland, J.B.; McMullen, M.D.; Buckler, E.S. 2008. Genetic design and statistical power of nested association mapping in maize. Genetics, 178: 539–551.

Zia, S.; Romano, G.; Spreer, W.; *et al.* 2012. Infrared thermal imaging as a rapid tool for identifying water-stress tolerant maize genotypes of different phenology. Journal of Agronomy and Crop Science, doi:10.1111/j.1439-037X.2012.00537.x.

8 Electrophoresis, Chromatography, and Mass Spectrometry

Thaís Regiani, Ilara Gabriela F. Budzinski,
Simone Guidetti-Gonzalez, Mônica T. Veneziano
Labate, Fernando Cotinguiba, Felipe G. Marques,
Fabrício E. Moraes, and Carlos Alberto Labate
Department of Genetics, University of São Paulo/ESALQ, Piracicaba, SP, Brazil

Introduction

The success in the improvements to plants during the 20th century is due, above all, to the use of natural variation, to the induction of genetic mutations, and to selection efficiency. The assessment and identification of genetic variants of interest, as well as the selection methodologies now in use, are based on phenotype evaluations (Pérez-de-Castro *et al*., 2012). Despite the vast number of better-quality plant species, very few have been targeted for studies involving the major leaders of systems biology: genomics, transcriptomics, proteomics, and metabolomics.

The first step to consider in a proteomic study is the harvesting and preparation of the samples. Meticulous care is essential in the preparation of high-purity samples in order to achieve quality results. The next step involves the protein separation processes: the protein fraction may be subjected to one-dimensional electrophoresis in polyacrylamide gel (SDS-PAGE) – which separates the proteins by molecular mass – or to two-dimensional electrophoresis in polyacrylamide gel (2D-PAGE) – which separates them by molecular mass and isoelectric point. The gel areas containing the protein (spots or bands) may be excised and subjected to a trypsinization procedure, which will yield a peptide mixture. The peptides derived from a given spot or band can then be analyzed by mass spectrometry (MS); a preceding separation may be carried out using liquid chromatography.

High-efficiency liquid chromatography combines the use of solvents (the mobile phase), a chromatographic column that usually contains

Omics in Plant Breeding, First Edition. Edited by Aluízio Borém and Roberto Fritsche-Neto.
© 2014 John Wiley & Sons, Inc. Published 2014 by John Wiley & Sons, Inc.

small-sized particles (the stationary phase), and high pressure. This powerful separation technique may be linked to mass spectrometry by means of high-precision, sophisticated equipment used in proteomics and metabolomics analyses.

The mass spectrometer consists of three compartments: an ion source, an analyzer, and a detector. The function of the ion source is to produce ions that are conveyed to the analyzer, where they are separated according to their mass/charge (m/z) ratio. Once separated, the ions proceed toward the detector wherein they collide and create an electric current which, in turn, generates a signal that is recorded, aggregated, and converted into a mass spectrum. The mass pattern obtained from the spectrum may then be analyzed with the aid of a database to identify the analytes.

With the recent advances of research in the fields of genomics and proteomics, another area that has been the subject of increasing interest since 2000 is metabolomics. The objective of metabolomics is to identify the metabolites involved in different biological processes related to the genotypic and phenotypic characteristics of a given individual.

As in proteomics, the crucial step in conducting metabolome-related studies is undoubtedly the preparation of a high-quality sample. The next step consists of analyzing such fractions by means of hyphenated techniques, which combine a separation method and an identification method. The main separation techniques used in metabolomics studies are gas (GC) and liquid (LC) chromatography coupled to instruments – such as those used in mass spectrometry (MS) or nuclear magnetic resonance (NMR) – that provide data for the structural identification of the metabolites. Again, as in proteomic studies, the data derived from mass spectrometry analyses, combined with the metabolite retention times, are recorded in online databases and analyzed with the aid of specific softwares.

Metabolomics is still viewed as a science in progress. Nevertheless, it is already a highly useful technique for medical diagnoses, as well as a method for differentiating genotypically distinct individuals in a population. It is also used in systems biology studies intended to provide a better understanding of the metabolic regulation and response to stressful conditions. Metabolomics has recently been used in the analysis of natural variance in cultivated plant species as part of a group of strategies to promote the genetic improvement of such plants.

In this context, the present chapter will describe the primary characteristics of the analytical separation, detection, and identification tools used in proteome- and metabolome-related studies.

Two-dimensional Electrophoresis (2DE)

After the proteins from the animal or plant tissue under investigation have been extracted, either totally or subcellularly, what remains is a complex sample composed of a mixture of proteins in different proportions. To be detected, quantified, and identified, these proteins must first be separated, because the accuracy in the identification of a peptide increases markedly as the complexity of the mixture diminishes.

One of the most common techniques to resolve complex protein mixtures is two-dimensional electrophoresis (2DE) due to its capacity to separate about 10 000 proteins. This technique, which in the past was synonymous with proteomics, is being replaced by more sophisticated methodologies. At present, however, 2DE is still widely used in proteomic analyses because of its low cost and reliable results. Two-dimensional electrophoresis can separate proteins on the basis of two characteristics: their isoelectric point and molecular mass.

Each protein has unique properties, such as size, molecular mass, amino acid sequence, and electric charge. Its electric charge is used to perform the first separation by 2DE with the aid of isoelectric focusing.

In this first separation step, the protein sample is placed on a polyacrylamide gel strip with an immobilized pH gradient. In the isoelectric focusing system, the charged strip (Figure 8.1a) is subjected to high voltage, which causes the protein to migrate into a pH gradient until its electric charge is reduced to zero and the protein rests on a definite area of the strip. The point that equates with a pH band where the positive and negative charges of a protein are at equilibrium is known as the isoelectric point. After focusing at the isoelectric point, the protein samples are taken through the second step of separation, by molecular mass difference, in which they undergo electrophoresis in polyacrylamide gel. The polyacrylamide concentrations can be adjusted to make the gel more or less porous after polymerization. The difference in porosity affects the speed with which proteins of different sizes will migrate into the gel.

The focused tape is treated with a buffer containing SDS (sodium dodecyl sulfate), a detergent that gives proteins a negative charge in proportion to their molecular mass. After this procedure, the tape is placed on top of the polymerized polyacrylamide gel and the electrophoresis is ready to begin.

When the polyacrylamide gel is exposed to high voltage (Figure 8.1b), the protein–SDS complexes detach from the isoelectric focusing tape and migrate through the gel at rates that depend on their ability to cross its pores. The larger the protein, the greater its difficulty of passing through the pores and, therefore, the higher the probability that it will remain on top of the gel.

Following the two-dimensional separation, the gel is first treated with a fixing solution to prevent the proteins from dispersing, and then with a staining solution to make them easier to detect and visualize for the subsequent analyses. Acetic acid is normally used to fix the proteins. For gel staining, there is a wider range of available possibilities; among them, silver and Coomassie blue are the preferred colorants. A two-dimensional profile of small roundish spots can be observed in the gels (Figure 8.1c): each spot represents a single protein or a group of several proteins.

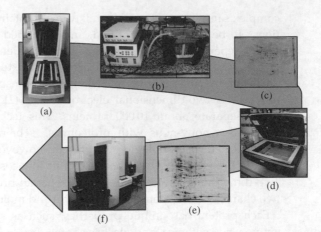

Figure 8.1   (a) Isoelectric focusing system; (b) electrophoresis system; (c) 2DE gel stained with Coomassie blue; (d) 2D gel image scanner; (e) 2D gel digital image; and (f) mass spectrometer. (See color figure in color plate section).

Next, the profiles of the stained gels should be recorded by means of an image digitization process to allow analysis of the protein spots. The electrophoretic profile images may be obtained with the aid of a special scanner (Figure 8.1d) that digitizes both sides of the gel to generate 3D images (Figure 8.1e) and allows the protein spots to be quantified by volume, intensity, and surface area. With the assistance of image-analysis software, spots from the same treatment may be grouped according to their location and distribution pattern on the gels, and compared with spots found in other treatments to detect differentially expressed proteins.

Once the images have been analyzed, the spots are excised from the gel and the proteins are eluted, enzymatically digested, and then analyzed for identification by mass spectrometry (Figure 8.1f).

With the use of this sequence of analytical methodologies by 2DE, weakly expressed proteins may be underestimated because the staining processes tend to highlight proteins that are found in higher concentrations in the proteic mix. There are other problems with the use of this sequence: some proteins are unsuitable for separation by isoelectric focusing, and the technique is not fast enough, particularly for the stages involving gel preparation, image analysis, and spot excision, which take considerable amounts of time when compared wiht other proteomic techniques.

Chromatography

Chromatography is an important physicochemical separation method. It is based on the principle of differential migration of the components in a mixture, which results from different interactions between two immiscible phases: a mobile phase and a stationary phase (Degani *et al.*, 1998). The great variety of combinations between the two phases makes chromatography a highly versatile technique that finds extensive application in different fields of science (Cass and Degani, 2001). As the components in the mixture have different levels of affinity for each of the phases, some compounds are retained in the stationary phase longer than others, which causes the molecules to separate. Component retention is determined not only by the physical and chemical properties of the two (mobile and stationary) phases, but also by experimental conditions (such as temperature and pressure) (Summerfield and Reid, 2010). Among the existing chromatographic procedures, liquid (LC) and gas (GC) chromatography are the methods of general preference for proteomic analyses as well as for metabolomics studies.

Liquid Chromatography

The name refers to the method in which the mobile phase is a liquid (Scott, 2003). It is used for the separation of compounds over the amount of time that they take to elute from the stationary phase (which is usually a separation column) with the help of the mobile phase. The separation occurs because the compounds have different affinities for the two phases. Compounds with a higher affinity for the mobile phase will elute quickly, while compounds with weaker affinities for the mobile phase will be retained in the stationary phase and be eluted later, over different periods of time; in other words, they have longer retention times (RT) (Allwood and Goodacre, 2010).

Liquid chromatography may be performed with paper, thin layer material, and, particularly, chromatographic columns associated with different types of detectors (diode array, ultraviolet, fluorescence, ammeter, and mass spectrometer) (Summerfield and Reid, 2010). Column chromatography emerged in the late 1960s with the use of small columns through which the mobile liquid phase was pumped under high pressure – a technique that was named high performance liquid chromatography (HPLC) (Dong, 2006). HPLC analyses require bottles for each mobile phase eluant (A and B), a pump, an injector, a separation column, and a detector. An aliquot of the sample is injected into the chromatograph (usually 5–25 μl) and separated in the column (Figure 8.2). The various components in the mixture migrate through the column at different speeds that result from the differences in partition between the mobile phase and the stationary phase. The constituents of the sample may be identified by comparing the experimental retention times with those obtained from known standards that have been analyzed previously under the same chromatographic conditions (Summerfield and Reid, 2010).

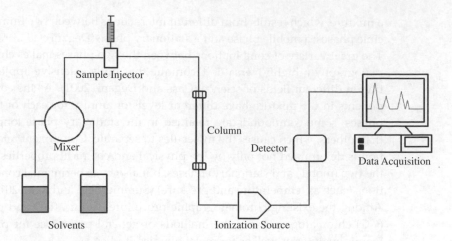

Sample Injector

Mixer

Solvents

Column

Detector

Data Acquisition

Ionization Source

Figure 8.2 Main components of a liquid chromatography system.

The solvents used in LC must be of high-purity grade and free of oxygen and other dissolved gases; the solvents must be filtered and degassed before use. The pump should provide the system with a continuous, pulse-free, highly reproducible flow rate to allow elution of the mobile phase at an adequate flow; it should also be programmable to create an appropriate solvent gradient during the separation (Summerfield and Reid, 2010). The columns for HPLC or UPLC (ultra performance liquid chromatography) analyses are generally made of stainless steel and measure between 10 and 30 cm in length and 3–5 mm in diameter. The stationary phase is chemically or physically bound to the internal material of the column and usually consists of 3 µm silica particles (Summerfield and Reid, 2010) (Figure 8.3).

In normal phase (NP) chromatography, the stationary phase is more polar than the mobile phase. Conversely, the mobile phase is the more polar one in reverse phase (RP) chromatography. Analytical separations are predominantly performed in RP due to the numerous advantages of this procedure, such as: the use of mobile phases, which are less costly and less toxic; RP columns are faster to equilibrate after a change of mobile phase; the possibility of conducting separations in the elution-by-gradient mode, which results in faster and better separations in the analyses; good repeatability of the retention times; a broad field of application because of the possibility of separating compounds with different polarities, molecular masses, functionalities, and so on (Summerfield and Reid, 2010).

The separation by RP is based on the coefficient of partition between the mobile polar phase and the stationary hydrophobic phase of the analyte. The mobile polar phase is generally composed of a mixture of methanol or acetonitrile in water. The first columns used in reverse phase

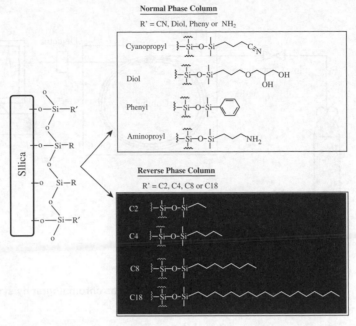

Figure 8.3 Diagram illustrating the most important stationary phases used in liquid chromatography. (Source: Modified from http://www.crq4.org.br/sms/files /file/hplc_araraquara_2012_site.pdf).

chromatography contained a stationary phase lined with nonpolar particles. These particles have since been replaced by molecular groups with permanent hydrophobic bonds, such as octadecyl (C18) groups attached to a silica base (Dong, 2006). This leads polar analytes to elute first, while nonpolar analytes interact strongly with C18 hydrophobic groups to form a "liquid phase" around the solid silica base.

Liquid chromatography is primarily used to separate strongly polar molecules that have high molecular mass and are not volatile (alcohols, phenols, and terpenes, for example).

Gas Chromatography

Gas chromatography (GC) is usually applied to the separation of compounds that can volatilize without decomposing; the separation is effected by a partition created between a mobile gas phase and a liquid or solid stationary phase inside an appropriate column.

The sample is injected into the inlet of the chromatographic column, vaporized, and then transported by a stream of inert gas through a packed or capillary column. In contrast to many other types of chromatography, the mobile phase does not interact with the analyte molecules: its only function is to carry it through the column. The components of the sample interact only with the stationary phase,

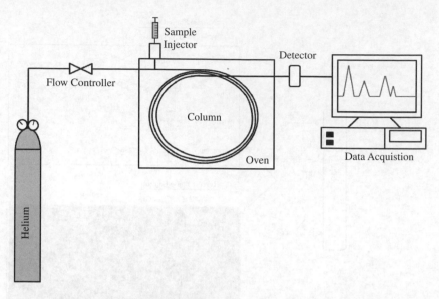

Figure 8.4   Main components of a gas chromatography system.

and the interactions are intermolecular (creating hydrogen bonds, for example) and hydrophobic (Collins *et al.*, 1997).

Briefly, gas chromatography is a method that involves first of all injecting the sample, which is vaporized before the carrier gas – usually helium, hydrogen, nitrogen, or argon – is injected into the column. As the gas flows through the column, the sample components are carried forward at speeds that are influenced by the degree of absorption, ion interchange, phase partition, and size of each component of the non-volatile stationary phase. Substances that interact strongly with the stationary phase remain longer in the column and, as a result, are separated from those that interact more weakly with the stationary phase. As they elute from the column, the substances are analyzed and/or quantified by a detector (Figure 8.4).

Owing to its simplicity, sensitivity and sturdiness, gas chromatography is one of the most important analytical techniques for mixtures of organic compounds with small molecular mass, boiling points up to 300 °C, and thermal stability; examples of such compounds are amino acids, sugars, organic acids, and esters. Sample groups that fail to meet these requirements may undergo an additional step, termed derivatization. In this case, the sample is treated with certain reagents, such as methoxyamine and *N*-methyl-*N*-(trimethylsilyl) trifluoroacetamide (MSTFA), to lower the boiling point of some of the organic compounds in the sample and allow it to be converted into a gas at lower temperatures.

One of the major characteristics of the gas chromatograph is that the gas phase column is located inside a chamber that allows the gas temperature to be controlled. Calibrated syringes are used to inject liquid samples through heated silicone diaphragms or septa placed at the inlet of the column; it is also possible to use automatic sampling injectors, which help to improve reproducibility and optimize analysis times. The sample intake port is maintained at about 50 °C, that is, above the boiling point of the least volatile compound in the sample (Skoog *et al.*, 2006).

The injected volume depends on the type of column: it can vary between 0.01 and 3 µl for capillary columns (0.25 mm in diameter), which are normally used to provide the best resolution in the case of complex samples. There are two injection alternatives for capillary columns: (i) *split injection*: the sample is divided at injection, and (ii) *splitless injection*: the sample is not divided at injection. The former procedure is preferable for analyses of highly concentrated solutions, because part of the sample is discarded after vaporization. Conversely, the splitless technique is generally the method of choice for the analysis of dilute solutions, because most of the vaporized sample contained in the injector is carried through the column. It should be pointed out that the latter method has poor reproducibility and, in addition, causes substantial wear in the column (Pereira and Neto, 2000).

Mass Spectrometry

Mass spectrometry is a powerful analytic technique that can be used to detect unknown compounds in very low concentrations and in chemically complex mixtures. This technique has played a fundamental role in the study of biochemical processes of proteins and metabolites, because it not only helps to detect and determine the structure of such molecules using fragmentation experiments, but it also makes it possible to quantify the molecules.

Since the late 1980s, with the development of an electrospray type ionization (ESI) source, this detector could be coupled with liquid chromatography. However, for metabolite analysis, there are now other ways of introducing a sample into a detector for mass spectrometry in addition to chromatographic separation. One of these ways is to inject complex samples directly, without resorting to previous chromatography separations.

Direct injection – also called direct infusion – is performed with the use of controlled flow rates by means of mechanical pumps that impel the piston of a glass or plastic syringe containing the sample. This type of injection has become widely used for MS analysis because it is fast, perfectly compatible with, and efficient for the analysis of numerous metabolites and for profiling samples (fingerprinting), provided that the detectors to be used ensure high resolution, accuracy, and

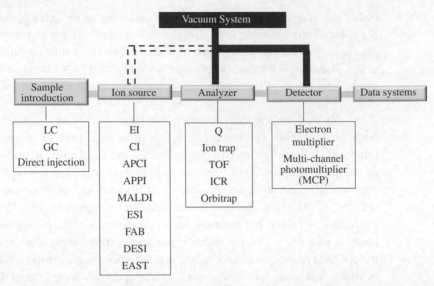

Figure 8.5   Components of a mass spectrometer. Introduction of the sample may be effected by liquid chromatography (LC), gas chromatography (GC), or even by direct injection with the aid of a mechanical pump. Some ion sources operate under vacuum, such as electron impact (EI) and chemical ionization (CI) sources. Examples of ionization sources that work under atmospheric pressure are as follows: chemical ionization (APCI), photoionization (APPI), matrix-assisted laser ionization and desorption (MALDI), electrospray ionization (ESI), and fast atom bombardment (FAB) sources, in addition to ionization sources that work at normal temperatures, such as desorption by electrospray (DESI), and sonic-spray ionization sources. The following analyzers may be cited as examples: quadrupole (Q) and ion trap analyzers, which have unitary and time-of-flight (TOF) resolution, ion cyclotron resonance (ICR), and the high-resolution Orbitrap. The most important detectors in use are the electron multiplier and the multi-channel photomultiplier (MCP) detectors.

reproducibility. Mass spectrometry is a very effective analytical technique when the purpose is to study the comparative profile of sample groups, such as those obtained from biological fluids, plant extracts or microorganisms.

The main components of a mass spectrometer are the ionization source, the analyzer, and the detector (Figure 8.5). Each of these components will be discussed briefly in the following paragraphs. An understanding of the function of each of these devices will shed light on their applicability and on the range of results that they provide.

Ionization Sources

In mass spectrometry, the function of the ionization source is to generate ions from the analytes of a sample and, in certain cases, to be able to transfer them to the gas phase. The possibility of using mass spectrometry for the analysis of biomolecules is due to breakthroughs that have been made, at the end of the 1980s, in the development of atmospheric pressure soft ion sources, such as electrospray ionization and matrix-assisted laser ionization (MALDI) devices (El-Aneed *et al.*, 2009).

The first ionization sources were electron ionization (EI) – which was also called electron impact – and chemical ionization (CI); both work under vacuum. In these cases, the molecular ions are formed with an excess of internal energy that causes them to fragment totally or partially. At present, these ionization sources are coupled to GC equipment for use in metabolite analysis. However, since proteins are polar, non-volatile, and thermally unstable macromolecules, an ionization by these methods may lead to their fragmentation (Banerjee and Mazumdar, 2012), which makes the analysis unfeasible, in this case.

In view of the limitations imposed by EI, other soft ion sources, which no longer require vacuum, have been developed to improve analyses by MS for molecules with higher molecular mass, such as peptides and proteins: MALDI was created in 1988 (Karas and Hillenkamp, 1988; Tanaka *et al.*, 1988) and ESI in 1989 (Fenn *et al.*, 1989). The ionization technique by MALDI and ESI for biological macromolecules was considered so important and innovative, that their designers, Koichi Tanaka and John Bennett Fenn, were awarded the Nobel Prize in Chemistry 2002 (The Nobel Prize in Chemistry, 2002).

The ESI technique is based on the injection of electrically charged droplets produced in solution by the addition of organic acids, such as formic acid. The usual function of this acid is to donate a proton to the analyte molecules. The resulting ions flow through an electrically charged capillary tube. The voltage difference in the capillary (normally 2–5 kV) induces charges to accumulate on the surface of the liquid at the discharge end of the tube, creating charged droplets. As these droplets cross the interface between the ion source and the intake port of the analyzer, they become smaller due to solvent evaporation caused by the high temperatures at the interface and by the constant flow of nitrogen through the interface. The droplets eventually shrink to a size at which the repulsion between ion charges is strong enough to explode the droplet, which leads the ion to be injected alone into the analyzer (Coulomb explosion) (Abonnenc *et al.*, 2010).

One of the best characteristics of ESI is its capability to ionize intact chemical species, such as proteins, with multiple electric charges (Fenn *et al.*, 1989), without fragmentation occurring during the ionization (Banerjee and Mazumdar, 2012).

The ESI-MS technique has outstanding features that include high sensitivity and selectivity, ease of use, and extensive applicability. Several classes of compounds can be analyzed by EMI, including proteins, nucleic acids, and even metal complexes, as long as the compounds are amenable to ionization or possess acid or basic sites (Heck and Van Den Heuvel, 2004). As the samples to be analyzed, must be introduced in solution, another advantage offered by ESI is the possibility to couple it, easily, with a liquid chromatographic technique, which will improve the separation and subsequent identification of the analytes. As a result, HPLC/ESI-MS has become a very powerful technique, which is capable of analyzing both low- and high-molecular mass molecules with different polarities in complex mixtures of biological samples.

The MALDI source is an ionization technique that was developed during the 1980s (Karas *et al.*, 1985; Tanaka *et al.*, 1988) and is used to detect and characterize biomolecules with molecular masses between 400 and 350 000 Da (Vlek *et al.*, 2012). One of the foremost uses of MALDI is to analyze gel spots in two-dimensional electrophoresis (2DE): after the enzymatic digestion *in gel*, the spots are ionized and analyzed. In most instruments that use MALDI as an ion source, the coupled mass analyzer is of the time-of-flight (TOF) type.

The principle of ionization by MALDI is based on the co-crystallization of the substance to be analyzed with a matrix in molar excess. This type of ionization yields charged ions in the gas phase by absorbing energy from the laser and transferring it to the analyte. The matrix is generally made of an organic aromatic acid that supplies ions to the ionization process of the sample. Since the nature of the matrix plays a fundamental part in the ionization of the compound, the choice of the matrix will depend on the objectives of the experiment. The most frequently used matrices are: α-cyano-4-hydroxycinnamic acid (HCCA); dihydroxybenzoic acid (DHB) for peptides (under 5 kDa) and lipids; and 3,5-dimethoxy-4-hydroxycinnamic acid (sinapinic acid) for proteins (above 5 kDa) (Mann *et al.*, 2001).

As in ESI, the ions generated by MALDI have low energy levels and create mostly monocharged species, which is helpful because it makes it easier to interpret mass spectra results, even for high-molecular mass compounds. There are other advantages in using ionization by MALDI instead of ESI; analyses with MALDI are faster, less costly, and the tolerance for contaminants, such as buffers and salts, is higher. However, MALDI is unfit for low-scale *m/z* analyses due to the very strong background noise from matrix ions.

Analyzers and Detectors

After the ionization, the molecules of the sample are transferred to an analyzer, where the ions are separated according to their mass/charge (*m/z*) ratios (all these procedures take place under vacuum). This can be done in several different ways, depending on the analyzer, which can be a quadrupole, an ion trap, or a TOF. Following the ion separation/selection step in the analyzer, the ions are driven forward to the detectors, which change the electric signal generated by the collision of each ion with the collector plate into a signal in the mass spectrum.

Quadrupole

The quadrupole-type analyzer consists of four metal rods arranged in parallel. As the gaseous ions generated in the ionization source reach the quadrupole, they are oriented by the parallel rods in response to the effect of the direct current (DC) and the radiofrequency (RF) applied to the rods. The voltage applied to the rods may be changed, which is quite convenient for the analysis because it allows the specific m/z ranges to be transmitted to the detector to be chosen. The rods function as filters, and are therefore more useful for quantitative analyses because of their ability to select an ion of interest in a highly sensitive and efficient manner. The quadrupole is an example of unit resolution analyzers.

One of the possible arrangements for this analyzer is the triple stage quadrupole, which consists of three quadrupoles disposed in tandem: the first quadrupole (Q1) selects the precursor ion, that is, it functions as a mass filter; in the second quadrupole (Q2), these ions are accelerated and fragmented by colliding with insert gas (argon, helium or nitrogen) molecules to generate so-called "offspring" ions; the last quadrupole (Q3) selects from among the offspring the one that is most characteristic of the precursor ion (Hopfgartner *et al.*, 2004).

The triple quadrupole operates in several modes, the prevalent mode being the ion production scan. In this case, Q1 functions as a mass filter, in which the voltage configurations are set to allow only ions with specific m/z values to cross it. These ions reach Q2, collide with inert gas molecules, and fragment. All the resulting fragments are analyzed in Q3, according to their m/z. This analyzer is used extensively with ESI. Selective monitoring of the reactions is also widely used in quantification methodologies, in which Q1 selects the desired ion for the study, Q2 fragments it, and Q3 selects the "offspring" ion that is characteristic of the ion under investigation. In the neutral loss scan mode, both Q1 and Q3 are programmed to select a neutral fragment and detect all the reactions that would lead to its loss. This mode requires: (i) both analyzers to conduct the scan simultaneously, (ii) a constant m/z difference to remain between them, and (iii) the selected reactions to be monitored (Sismotto *et al.*, 2013).

The use of quadrupoles offers the following advantages: they provide good reproducibility, they are suitable for quantification experiments, and they are very sensitive. Their resolution, however, is limited.

Ion Trap

The expression *ion trap* literally defines how this type of analyzer works: it traps ions. These analyzers accumulate ions inside them and operate by adjusting alternate currents and radiofrequencies simultaneously. The ion capture procedure allows a gradual release of ions in the desired m/z range, resulting in controlled separations, which, in turn, provide high-sensitivity spectra. It is also used in sequential MS experiments with $n > 2$ fragmentation (MS^n) for the study of the fragmentation mechanism of relatively complex molecules – for example, for a structural elucidation to determine the position of the double bonds in unsaturated lipid chains, or for the elucidation of a peptide chain.

One of the advantages of the ion trap lies in its high sensitivity, which derives from the fact that the ions are fragmented and pre-concentrated in the same site, at different times, and then analyzed. This analyzer also allows conducting sequential MS^n experiments over time. An ion trap has, however, a few disadvantages to be considered, such as the impossibility of alternating acquisitions between the product ("offspring") ion scan mode and the neutral loss scan mode. These analyzers are not powerful enough to analyze more complex matrices, and they provide low reproducibility. Furthermore, their resolution is unitary and their dynamic range is too narrow for quantification (Sismotto *et al.*, 2013).

Time-of-Flight (TOF)

Time-of-flight is one of the principal analyzers for proteins. The analyzer was given this name to reflect its ability to measure the m/z ratio of ions by estimating the time they take to travel through the flight tube with an acceleration provided by constant voltage; this establishes a proportion between the time spent and the m/z ratio (Cardoso *et al.*, 2001). Ions with lower molecular masses reach the detector more quickly than ions with higher molecular masses. The time-of-flight tube operates under vacuum and without an electric field.

This analyzer is classed as a high-resolution instrument because it can separate analytes with a precision of up to four digits after the decimal point. This level of resolution is due to the fact that it contains an electrostatic reflector (reflectron), which increases the optical path that the ions must cross, thus lengthening the time-of-flight and enhancing the resolution of the equipment. The use of the reflector also corrects the effects of small differences in the kinetic energy initially received by ions with the same m/z ratio, which cause differences in their velocity. This electrostatic reflector is composed of an array of grids that create an electric field, which delays the flight of the ions and reflects them back to the flight tube in a curved path. While traversing this curve, ions that have the same m/z ratio, but higher kinetic energy and velocity levels, will penetrate deeper into the reflector and take longer to leave it. Conversely, slower ions exit the reflector more quickly, thereby compensating for the difference in velocity between the two species.

One of the advantages of using TOF analyzers is that, as with MALDI, they are connected to pulsed ionization sources that shorten the time of analysis. TOF analyzers also provide high ion transmission and high sensitivity.

Hybrid Analyzers: Quadrupole–Time-of-Flight

The use of two analyzers in sequence, such as a quadrupole and a TOF (Q–TOF) analyzer, results in a versatile combination of great importance for the study of biomolecules. The hybrid Q–TOF analyzer combines: (1) the quadrupole, which selects the ion; (2) a collision cell (filled with inert gas), which fragments the ion selected in the quadrupole; and (3) the TOF tube, a high-resolution, high-accuracy, high-speed spectral-acquisition analyzer, which separates the fragmented ions (Stolker *et al.*, 2004). These analyzers are usually associated with ionization sources of the ESI or MALDI types.

Ion Detectors

After crossing the analyzer, the ions proceed to the detector, inside which they collide and thereby generate an electric current. The amplification of these collisions is transformed into a signal that will be recorded, aggregated, and converted into a mass spectrum.

An example of such a detector is the electron multiplier, which is based on the principle of the Faraday cup. This type of detector uses a serial array of collector plates maintained at increasing potential levels, resulting in a cascade of electrons with a typical amplification gain on the order of 1 000 000 times.

Another option is to use photomultiplier plates. In this case, when the ions collide with the plates, they issue electrons that collide with a phosphorus screen and eject photons, which are then detected in cascade. The emitted photons remain inside a system that is isolated from the rest of the equipment to avoid the presence of other particles that might interfere with the detection. This arrangement increases the sensitivity and lifespan of the plate.

Data Analysis

All mass spectrometers include proprietary programs both for the acquisition and the processing of data. These programs are usually designed with a friendly and intuitive interface. However, when a more detailed analysis of the data is required, some of the programs may fail to provide all the appropriate resources, particularly because most programs have closed source codes to protect the copyright of their developers. This makes it difficult to change existing analytical procedures or implement new ones.

Another problem is the format used in data exit files. Each company has its own file format. Proprietary programs process and analyze only data generated by certain machines.

With the increasing use of mass spectrometry for proteomics and metabolomics analyses, several research groups have been developing free commercial programs, not only for the extraction of data, but also for the analysis of peptides and metabolites. Some of the better known commercial programs are: SIMCA (Umetrics, Umeå, Sweden), MarkerLynx (Waters, Milford, MA, USA), ProteinLynx (Waters, Milford, MA, USA), MarkerView (Applied Biosystems, Foster City, CA, USA), MassHunter (Agilent Technologies, Santa Clara, CA, USA), Rosetta Elucidator (Rosetta Bio Software, Seattle, WA, USA), MetAlign (Plant Research International B.V., Wageningen, The Netherlands), Scaffold (Proteome Software, Portland, OR, USA), and Mascot (Matrix Science, London, UK), among others.

Some of the better known free access softwares are: XCMS (Smith *et al.*, 2006), Mzmine 2 (Pluskal *et al.*, 2010), PeakML/mzMatch (Scheltema *et al.*, 2011), Ideom (Creek *et al.*, 2012), and Mascot (Matrix Science, London, UK), among others.

The processing of raw data usually constitutes the most time-consuming stage for a researcher, since the correct interpretation of the results is known to pose substantial challenges. The most common procedures for the analysis of such data include removal of noise from chromatograms, deconvolution of spectra, detection and integration of peaks, alignment of chromatograms, identification of peptides and metabolites, and, if applicable, quantification of the compounds (Shulaev, 2006).

The identification of proteins is conducted by associating the information produced by a peptide full-scan analysis and its respective fragmentation pattern, using algorithms from analysis programs (Mascot and ProteinLynx, for example). These data should be combined to reconstitute the peptides – a necessary step for the identification of the proteins. To do this, the peptides are compared with proteic sequences available in databases such as NCBI (http://www.ncbi.nlm.nih.gov/) and Swiss-Prot (http://www.uniprot.org), or specific databases for the organism being studied. As an outcome of this analysis, predicted proteins are identified and associated with scores and e-values.

Metabolite identification, on the other hand, is associated with the type of chromatography used during the analysis. When the metabolites are to be identified by GC-MS, it is possible to use a range of different information, such as that related to the derivatization of the metabolites and their characteristic fragments, as well as their retention times and indices (Tohge and Fernie, 2009). On the other hand, when the LC-MS technique is employed, metabolites identification is still a serious challenge (Creek *et al.*, 2012), because it is complicated in particular by the numerous variables that need to be adjusted in the instruments.

For these reasons, peak identification generally requires the use of complementary strategies, which involves considering resources such as information from already published data and available standards. These factors make it difficult to build an universal database for the identification of data acquired from LC-MS.

The databases used in peptide and metabolite identification should be selected according to the purpose and type of the samples being analyzed (Tohge and Fernie, 2009). Thus, the leading characteristics available in the databases, which are generally sought for the comparison of peaks, are: m/z values, retention times and indices, e-values, scores, and fragmentation spectra. Any previous knowledge regarding the samples to be analyzed should also be applied, since there are several databases grounded on the literature, in which the data are organized according to biosynthetic, organism-specific pathways for example. Therefore, previous understanding of an experiment will result in easier, more accurate analyses.

Among the best-known databases, the following stand out: KEGG (Kanehisa *et al.*, 2006), PlantCyc (http://www.plantcyc.org/), MetaCyc (Caspi *et al.*, 2012), MassBank (2006) (http://www.massbank.jp/), Metlin (http://metlin.scripps.edu/), Golm Metabolome Database (Kopka *et al.*, 2005), UniProt (http://www.uniprot.org/) and NCBI.

The statistical analysis of proteomics data may be performed with different tests such as the student t-test and ANOVA, among others. The choice of the statistical test to be used should be based on the adopted analytical strategy, such as the proteic profile of a sample or a comparative analysis between groups of samples. These analyses not only produce data for preferentially and differentially expressed proteins, which are a determinant of a given functional category, but they also allow the construction of networks of interaction between the identified proteins and their functions, by means of programs such as Blast2GO (http://www.blast2go.com/b2ghome) and Cytoscape (http://www.cytoscape.org/), among others.

In the case of metabolomics data, a typical data analyses method used is the principal component analysis (PCA), in which a set of correlated variables is transformed into a smaller number of uncorrelated variables that constitute the principal components. This type of analysis allows identification of groups, tendencies, and outliers in the analyzed samples (Hagel and Facchini, 2007).

On the basis of the information provided here, it is clearly of paramount importance for the researcher to be thoroughly familiar with the organism being studied, as well as with the available analytical and statistical tools that need to be deployed for the optimal acquisition and interpretation of the results (Figure 8.6).

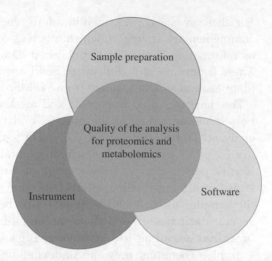

Figure 8.6 The quality of the proteomics and metabolomics analysis is dependent on the sample preparation, on the instrument used for analysis, and on the software used for interpretation of the data. (Source: Adapted from Thermo Scientific Pierce catalog for Mass Spectrometry Sample Preparation –V2, 2010).

References

Abonnenc, M.; Qiao, L.A.; Liu, B.H.; Girault, H.H. 2010. Electrochemical aspects of electrospray and laser desorption/ionization for mass spectrometry. Annual Review of Analytical Chemistry, 3: 231–254.

Allwood, J.W.; Goodacre, R. 2010. An introduction to liquid chromatography–mass spectrometry instrumentation applied in plant metabolomic analyses. Photochemical Analysis, 21: 33–47.

Banerjee, S.; Mazumdar, S. 2012. Electrospray ionization mass spectrometry: A technique to access the information beyond the molecular weight of the analyte. International Journal of Analytical Chemistry, 2012: Art. ID 282574, 40 pp.

Cardoso, A.S.; Pontes, F.C.; Souza; G.G.B., Mundim, M.S.P. 2001. Um espectrômetro de massas de tempo-de-voo para o estudo da ionização de amostras gasosas por elétrons rápidos (0.5 – 1.0 kEV). Química Nova, 24: 315–319.

Caspi, R.; Foerster, H.; Fulcher, C.A.; *et al*. 2012. The MetaCyc database of metabolic pathways and enzymes and the BioCyc collection of pathway/genome databases. Nucleic Acids Research, 40: 742–753.

Cass, Q.B.; Degani A.L.G. 2001. Desenvolvimento de Métodos por HPLC: Fundamentos, Estratégias e Validação. 1st edn. São Carlos: UFSCar, 77 pp.

Collins, C.H.; Braga, G.L.; Bonato, P.S. 1997. Introdução a Métodos Cromatográficos, 7th edn. Campinas:UNICAMP.

Creek, D.J.; Jankevics, A.; Burgess, K.E.V; *et al*. 2012. IDEOM: an Excel interface for analysis of LC-MS-based metabolomics data. Bioinformatics, 28 (7): 1048–1049.

Degani, A.L.G.; Cass, Q.B.; Vieira, P.C. 1998. Cromatografia: um breve ensaio. Química Nova Escola, (7): 21–25.

Dong, M.W. 2006. Modern HPLC for Practicing Scientists: Introduction. 1st edn. Hoboken: Wiley-Interscience, 286 pp.

El-Aneed, A.; Cohen, A.; Banoub, J. 2009. Mass spectrometry, review of the basics: electrospray, MALDI, and commonly used mass analyzers. Applied Spectroscopy Reviews, 44: 210–230.

Fenn, J.B.; Mann, M.; Meng, C.K.; *et al.* 1989. Electrospray ionization for mass spectrometry of large biomolecules. Science, 246: 64–71.

Hagel, J.M.; Facchini, P.J. 2007. Plant metabolomics: analytical platforms and integration with functional genomics. Phytochemistry Reviews, 7 (3): 479–497.

Heck, A.J.R.; Van Den Heuvel, R.H.H. 2004. Investigation of intact protein complexes by mass spectrometry. Mass Spectrometry Review, 23: 368–389.

Hopfgartner, G.; Varesio, E.; Tschäppäet. V.; *et al.* 2004. Triple quadrupole linear ion trap mass spectrometer for the analysis of small molecules and macromolecules. Journal of Mass Spectrometry, 39 (8): 845–855.

Kanehisa, M.; Goto, S.; Hattori, M. *et al.* 2006. From genomics to chemical genomics: new developments in KEGG. Nucleic Acids Research, 34: D354–D357.

Karas, M.; Bachmann, D.; Hillenkamp, F. 1985. Influence of the wavelength in high-irradiance ultraviolet laser desorption mass spectrometry of organic molecules. Analytical Chemistry, 57: 2935–2939.

Karas, M.; Hillenkamp, F. 1988. Laser desorption ionization of proteins with molecular masses exceeding 10 000 Daltons. Analytical Chemistry, 60: 2299–2301.

Kopka, J.; Schauer, N.; Krueger, S.; Birkemeyer, C.; *et al.* 2005. GMD@CSB.DB: the GolmMetabolome Database. Bioinformatics, 21 (8): 1635–1638.

Mann, M.; Hendrickson, R.C.; Pandey, A. 2001. Anlysis of proteins and proteomes by mass spectrometry. Annual Review Biochemistry, 70: 437–473.

Pereira, A.S.; Neto F.R.A. 2000. Estado da arte da chromatography gasosa de alta resolução e alta temperatura. Química Nova, 23: 370–379.

Pérez-De-Castro, A.M.; Vilanova, S.; Cañizares, J.; *et al.* 2012. Application of Genomic Tools in Plant Breeding. Current Genomics, 13 (3): 179–195.

Pluskal, T.; Castillo, S.; Villar-Briones, A.; Oresic, M. 2010. MZmine 2: Modular framework for processing, visualizing, and analyzing mass spectrometry-based molecular profile data. BMC Bioinformatics, 11: 395.

Scheltema, R.A; Jankevics, A.; Jansen, R.C.; *et al.* 2011. PeakML/mzMatch: a file format, Java library, R library, and tool-chain for mass spectrometry data analysis. Analytical Chemistry, 83 (7): 2786–2793.

Scott, R.P.W. Principles and practice of chromatography. 2003. In: Scott, R.P.W.1. (ed.) Chrom-ed book series: Book 1, http://www.library4science.com (accessed 25 February 2013).

Shulaev, V. 2006. Metabolomics technology and bioinformatics. Briefings in Bioinformatics, 7 (2): 128–139.

Sismotto, M.; Paschoal, J.A.R.; Reyes, F.G.R. 2013. Aspectos analíticos e regulatórios na determinação de resíduos de macrolídeos em alimentos de origem animal por cromatografia líquida associada à espectrometria de massas. Química Nova, 36 (3): 449–461.

Skoog, D.A.; West, D.M.; Holler, F.J.; Crouch, S.R. 2006. Fundamentos de Química Analítica. 8th edn. São Paulo-SP, Brazil: North American Thomson Publisher Version.

Smith, C.A; Want, E.J.; O'Maille, G.; *et al*. 2006. XCMS: Processing mass spectrometry data for metabolite profiling using nonlinear peak alignment, matching, and identification. Analytical Chemistry, 78 (3): 779–787.

Stolker, A.A.M.; Niesig, W.; Fuchs, R.; *et al*. 2004. Liquid chromatography with triple-quadrupole and quadrupole-time-of-flight mass spectrometry for the determination of micro-constituents – a comparison. Analytical and Bioanalytical Chemistry, 378: 1754–1761.

Summerfield, S.; Reid, H. 2010. Introduction to Chromatography: Chromatography. 1st edn. Loughborough: Loughborough University.

Tanaka, K.; Waki, H.; Ido, Y.; *et al*. 1988. Protein and polymer analyses up to *m/z* 100 000 by laser ionization time-of flight mass spectrometry. Rapid Communications in Mass Spectrometry, 2: 151–153.

The Nobel Prize in Chemistry. 2002. Nobelprize.org. http://www.nobelprize.org/nobel_prizes/chemistry/laureates/2002/ (accessed 7 February 2013).

Tohge, T.; Fernie, A.R. 2009. Web-based resources for mass-spectrometry-based metabolomics: a user's guide. Phytochemistry, 70 (4): 450–456.

Vlek, A.L.M.; Bonten, M.J.M.; Boel, C.H.E. 2012. Direct matrix-assisted laser desorption ionization time-of-flight mass spectrometry improves appropriateness of antibiotic treatment of bacteremia. PLoS ONE, 7 (3): e32589; doi: 10.1371/journal.pone.0032589.

9 Bioinformatics

J. Miguel Ortega[a] and Fabrício R. Santos[b]

[a]Department of Biochemistry and Immunology, Federal University of Minas Gerais, Belo Horizonte, MG, Brazil
[b]Department of General Biology, Federal University of Minas Gerais, Belo Horizonte, MG, Brazil

Introduction

There has been a big revolution in the generation of biological data since the beginning of the human genome project in the 1990s. Up till then, data could be stored in the hard disk of a personal computer. In the last two decades there has been a growing demand for virtual space for the processing of these data. Moreover, to follow this exponential increase in volume of molecular data, there has also been the need for increased computational capacity for storage, processing, and analysis (Prosdocimi and Santos, 2004). In this context, great demands were imposed on the development of this new "Science," for example, the development of computational platforms suitable for the implementation of bioinformatics analysis of molecular data. The limitations include both hardware and software components. However, various computing and analysis tools have been developed to deal with this massive amount of data arising from genomics, proteomics, metabolomics, metagenomics, and so on.

In this chapter, we describe the history and current state of bioinformatics as applied to the processing of large amounts of data generated by new "omics."

The "Omics" Megadata and Bioinformatics

Data from biological systems are relatively complex compared with that from other scientific areas, given their diversity and their interrelationship, as demonstrated by the results generated by genomics projects (Figure 9.1). According to the fundamental knowledge of the genome assembled from DNA sequences, the goal is to understand the complexity of the whole organism; for instance, which genes are related

Omics in Plant Breeding, First Edition. Edited by Aluízio Borém and Roberto Fritsche-Neto.
© 2014 John Wiley & Sons, Inc. Published 2014 by John Wiley & Sons, Inc.

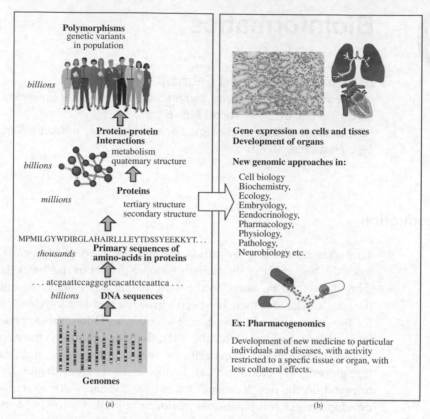

Figure 9.1   Biological data increase (a) and practical applications of the genomic knowledge (b). (See color figure in color plate section).

to drug response, one of the goals of pharmacogenomics. However, at the moment, this is only partially possible, due to the great complexity of data and theoretical limitations, leading to a great demand for bioinformatics. Firstly, one seeks to understand the molecular structures of proteins and other gene products, such as the functional RNAs, the interactions between several of these molecules synthesized from the genome, as well as those with other structural and functional biological molecules (DNA, carbohydrates, lipids, etc.), the various cellular metabolic pathways and the role of genetic variability represented by various forms of each gene product. All this information provided by genomic science is only possible to be organized, analyzed, and interpreted with the support of bioinformatics.

Currently, bioinformatics is essential for handling any type of biological data, especially the "omics" megadata. Bioinformatics can be defined as an approach which covers all aspects of acquisition, processing, storage, distribution, analysis, and interpretation of biological information. Through the combination of procedures and techniques of mathematics, statistics, and computer science, several tools can be called upon that help us understand the biological meaning contained in "omics" biological data. Furthermore, by creating databases with information already processed, bioinformatics accelerates the investigation in other biological areas, such as medicine, biotechnology, agronomy, and so on.

Hardware for Modern Bioinformatics

The bioinformatics tools have characteristic features, but a general trend has appeared in recent years: the change from working with software installed on your own computer to the use of a server shared by other research groups. More recently, research is now conducted on large institutional servers. Along with this, the adjustment of researchers to the Linux environment became mandatory. Nowadays it is common for undergraduate students in the biological sciences area to access remote servers. The connection is almost always done by means of a program that runs the connection protocol *SSH* (Secure Shell) and it is interesting that the triggering of some programs, the monitoring of workflows or even displaying results can be done with applications installed on tablets or even smartphones!

In Brazil, the National System of High Performance Computing (SINAPAD) in particular is becoming an increasingly important resource used by research groups in bioinformatics, and it is quite simple to get an account for project development on the servers of the centers that comprise the SINAPAD, which are known as CENAPAD (National Center of High Performance Computing). Typically, each user runs programs that occupy a significant fraction of the machine, say about 50–100 computational cores, which would not be feasible on a local server, accomplishing its outcome in about a day to a week, depending on the amount of data and analysis. If more cores are available in the machine, parsimoniously it is possible to use them to accelerate the processing of the required routines.

Recently, institutions began to acquire servers that characteristically have high quantities of memory addressed by each computer core, the so called shared memory architecture, often reaching 2 TB (terabytes). This facilitates the assembly of large and complex genomes such as *brazil wood* (a tree) and the Amazonian manatee (a mammal), not yet sequenced (estimated at 3.8 and 4.6 billion bases, respectively) because the assembly requires that too many small sequences, generated in large quantities by new generation sequencers, are associated with each other, therefore requiring the addressing of a large amount of memory. Thus, the process can be done in a single step, because the various mounting possibilities can be tested simultaneously. Shared memory machines of about 2 TB are becoming accessible in several institutions, including the SINAPAD. The typical storage in hard drives on these servers is around 100 TB.

Thus, today bioinformaticians do not work in their own computers, but send the data to be processed remotely in hardware with the computational cluster format. Invariably, these servers operate various Linux operating systems distributions, such as *RedHat Enterprise*, *CentOS* or *Suse*. In addition, some also count with queue management systems to distribute the tasks triggered by users along several computational cores, such as the *OpenPBS* and *SLURM*. Many current analysis programs will have some version that allows for parallelism, which is when multiple computing cores are triggered at the same time to run routines whose results are gathered by other computational cores. In these cases, one can use a parallelism controller, such as *OpenMPI*, to make the work easier. Task requests on servers are usually made by the *SSH* protocol mentioned previously, but there is a great tendency to migrate to what is called a web-service. This is a new protocol that allows for the use of rented computing resources or "cloud computing." This is a good choice for researchers who do not have access to high-performance servers and do not require high performance computing constantly. Currently, it is typical of bioinformaticians to manage small-to-medium web servers, to present the results by means of dynamic pages, which access databases to perform compilations made at the time, answering the queries made by other researchers interested in the data.

Software for Genomic Sequencing

Today bioinformatics is a science that deals with the exploitation of existing information in living organisms, in biological systems. Often, the first step to explore this information is the processing of data from genomic projects.

The first automated sequencers provided data readings in the region of hundreds of bases in size (typically 500–1000). A detail overlooked in these early days of genomics was that the sequencers did not generate DNA sequences, but a set of fluorescence peaks, usually interpreted by the program *Phred* (http://www.phrap.org/phredphrap). The molecular weight of the fluorophore in the bases that were linked to dideoxyriboses, which disrupt the polymerization in the Sanger method, is differentiated for each base. Thus, software is needed to verify and to edit the position of the fluorescence peak. More importantly, the *Phred* program estimated, through examining the format of the peak, the probability (quality) of the correct determination of the corresponding base, therefore expressing the accuracy of the base calling.

Since that time, DNA sequences have been stored together with their probability of error. Copying what had been used to express the concentration of protons as pH, software determines "-log error chance," so a chance of error of one in 10 000, that is 10^{-4}, becomes 4. However, to avoid saving a possible decimal point to the hard drive, it is multiplied by ten, so the quality value 40 refers to a small chance of error in determining the base of 1/10 000, that is, 0.01%. This *Phred* value is known as the "quality" value of the determination of the base (base calling). All automatic capillary sequencers based on the Sanger method provide a *Phred* or equivalent quality value and, sometimes with some small difference in the calculation, the next generation sequencers also express the chance of error in determining the base.

The chance of error of 0.01% (i.e., 99.99% certainty) had been stipulated as the minimum precision value in the sequencing of the genomes of the first model organisms, such as yeast, the fruit fly, and man. In bioinformatics, it was said that the DNA should be sequenced until all the bases attained *Phred* quality value equal to or greater than 40. Therefore, it is understood that bioinformatics participates from the first step of the advancement of the knowledge of genomes. Currently, it is not uncommon for bioinformaticians to choose to work with chances of error greater than 0.01% on certain projects, because invariably the reads become longer, since the chance of error increases at the tail, where the accuracy of any method tends to decrease in almost all techniques. More recently, to save disk space, each value of quality has been encoded by a character and it is customary to accommodate sequence and quality in a single file. Thus, the quality values 10 and 20 are saved as + and 5, respectively (Figure 9.2 – comparing formats *Phred* and *FASTQ*). To store alignments of sequences to a reference one, there is the *SAM* (Sequence Alignment/Map) format and its derived version, the *BAM* format, which is easier to parse.

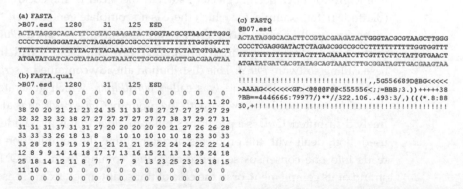

Figure 9.2 DNA sequence in FASTA format (a) and quality file FASTA.qual (b), both generated by *Phred* software, compared with *FASTQ* format (c). *FASTQ* files incorporate the sequence of bases identified by the "@," separated by the symbol "+" from quality data encoded by different characters, for example, the characters ".," "+," and "5" match the *Phred* values 0, 10, and 20, respectively. Regions of low quality (bases in red) had the quality values set to zero and will be dropped from the processed sequence. (See color figure in color plate section).

Software for *Contig* Assembling

The process of assembling large DNA sequences initially starts with the search for similar regions that drive the generation of clusters of sequences linked by these overlapped regions, which we call *contigs*. These, in turn, can be reconnected in increasingly larger *contigs*, until they form a chromosome or a complete genome. At first glance, it might seem that the junction of sequences exported by sequencers is only needed when it comes to determining the full sequence of genomes. For RNA sequencing, DNA copies which are made of RNA with the enzyme reverse transcriptase (cDNA) are also partially sequenced and need to undergo an assemblage to generate the continuous sequence of that RNA. Thus, sequence assembling is as useful in transcriptomics as in genomics.

In transcriptomics, assembling also produces the count of how many transcripts are found for each gene, that is, it expresses the abundance of transcripts. In cases where the genome of the organism has already been published, this work is simplified, because, instead of working with the assembly from scratch, one can anchor the transcripts to the published genome. When the genome is not available, it is possible to use a reference genome, which is the term used to define a genome that is closely related, often from the same genus of the organism of interest, or even the same species. The trick of anchoring small sequences in reference genomes is also commonly used to assemble novel genomes, which have great similarity to some already available. This brings us to the strategic importance of multiple genomes being available, although some taxonomic groups have been neglected, such as the order of the cockroach (Blattodea), for example, for which there is no complete genome yet.

The first well-known software for *contig* assembling is called *Phrap* (http://www.phrap.org/phredphrap), and is distributed along with the base calling software *Phred*. This distribution already contained a *script* that automated analysis and was called *phredPhrap*, which generated the *contigs* that could be viewed with the program *Consed*, which is also freely distributed. *Cap3* is a program similar to *Phrap* that is also widely used. Both dealt with the problems of determining the combination of reads into one consensus sequence or *contig*, coming from one DNA strand or its complement, or in the case of transcriptomics, from "sense" or "anti-sense" strands. These programs join the sequences through identical overlapping regions assembling two or more independent reads into just one, a consensus sequence. Note that the base of the assembly is the alignment of individual sequences, a technique widely used in bioinformatics that was used for the development of several assembling programs.

Assembly Using the Graph Theory

Only recently have the next generation DNA sequencers begun producing individual sequences of hundreds of bases. Until last decade, the readings were only tens of nucleotides, which led to the development of software for assembly of *contigs* through a different approach, because it was impossible to deal with alignments that overlap so slightly. However, in compensation, these sequencers can generate several million sequences in a single run. Moreover, together with the small overlap between them, it is not possible to generate contigs by comparing all sequences against all based on pairwise alignment. Fortunately, computing science often provides solutions in advance that can be applied to new problems. In this case, it is the Graph Theory. This methodology deals with chains of elements forming networks to address complex problems, such as determining the best route for spreading phone signals through subsequent antennas, and it is also used in the internet to find a machine through its IP address considering the most parsimonious path.

A widely used software for genome assembly of readings obtained with the modern sequencers is *Velvet* (http://www.ebi.ac.uk/~zerbino /velvet), based on Graph Theory. A window of a few bases walks through each one of the millions of small sequences, and by verifying that it can be connected with other short sequences, the program will connect them forming a huge network. At the end, the goal is simply to determine the network path (Bruijn Graph) that returns a *contig* or complete genome (Figure 9.3). While one can imagine that repetitive DNA might cause a bifurcation in the thread because the reading has several supposed continuities in different regions, fortunately Graph Theory deals with this and presents techniques to determine the appropriate path, solving the problem globally. For example, when there is a fork

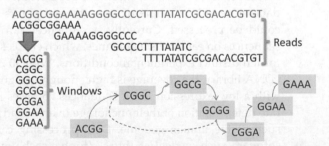

Figure 9.3 Assembly using the Graph Theory. The readings produced by the sequence are split into windows and a network is formed, much more complex than that shown above. The algorithm next seeks to deduce a path passing through all nodes of the network only once (the dashed path is eliminated). The basic procedure to solve this problem already existed, before being used in genome assembly. (See color figure in color plate section).

in the reading, the correct path to follow is the branch of the fork that does not end abruptly. Of course, the use of Graph Theory for dealing simultaneously with millions of small reads requires the use of a lot of RAM memory, typically 100 GB for transcriptome analysis and bacterial genomes and at least 2 TB for the genomes of animals and plants.

New Approaches in Bioinformatics for DNA and RNA Sequencing

Current technologies make use of the same sequencing strategy as that used in sequencing the human genome: the determination of the sequence of two ends of DNA fragments of known size. One of the first genome sequencing initiatives, performed by the company Celera, used readings of about 500 bases of sequences derived from three known sizes, 2, 10, and 50 kb, and the assembly was held in a computer that could address the "impressive" 4 GB of RAM memory. Currently, in addition to such an approach, known as *paired ends*, with the use of new methodological strategies, a chemical group is added to the end of a molecule of, for example 10 kb, and after circularization the molecule is fragmented, and the chemical group previously added is biochemically captured. Now, it is possible to sequence this fragment, which contains the information for both ends of the long initial molecule. This methodology is known as *mated pair*. The assembly software takes advantage of the information that in about 10 kb, the sequence B downstream of sequence A should be found in the network. The process, as stated earlier, is facilitated when the small new readings generated by the new sequencers can be anchored to a reference genome.

The use of new generation sequencing equipment in studies of transcriptomics, a technique known as *RNAseq*, is replacing the use of microarray methods, due to the great simplicity of implementation. Moreover, there was often enormous pressure on those who collected the data, since only a couple of experimental points (control and treated) could be processed. Currently, a sequencer can produce 150 million sequences on each of its eight lanes, which enables analyses in triplicate for ten distinct experimental conditions. With 40 million sequences per cDNA library, the coverage is high enough to significantly detect genes with a low expression level. Sequences 75 bases long suffice for accurate determination of their anchorage in an already sequenced genome, such as the human genome. Data processing can be done with free software such as *TopHat* (http://tophat.cbcb.umd.edu) and *Cufflinks* (http://cufflinks.cbcb.umd.edu), tools for analyzing *RNAseq*, identifying new genes and *splicing* variants, as well as determining differential expression (Trapnell *et al.*, 2012) (Figure 9.4). Nevertheless, as occurred during the epoch of the domain of the microarray, the determination of differentially expressed genes remains a big challenge for the various

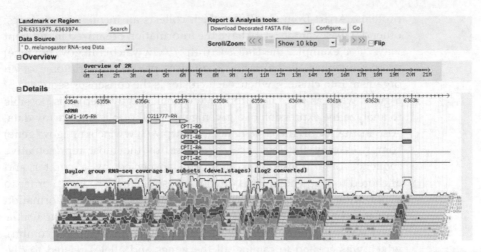

Figure 9.4 Expression profiles based on *RNAseq*, extracted from the database *FlyBase* (http://flybase.org), with the representation of the genome (upper scale), transcripts (*exons* represented by boxes and *introns* by lines), and the coverage of different regions by the anchored sequences, generated by next generation sequencing equipment. (See color figure in color plate section).

types of software available, especially when the expression is low. This invariably leads the researcher to confirm, through subsequent analyses, if the difference is real or whether it is a false positive (Soneson *et al.*, 2013).

Databases, Identification of Homologous Sequences and Functional Annotation

Owing to the immense amount of data generated in numerous laboratories around the world, it is necessary to arrange them in an accessible manner, so as to avoid redundancy for scientific research, and to enable analyses to be made by the maximum number of scientists. The construction of databases to store information of DNA sequences and whole genomes, of proteins and their three-dimensional structures, and processed information of protein interaction networks, metabolomics, as well as several other complex results from different "omics," has been a big and extremely important challenge.

One of the first biological databases was developed by the National Center for Biotechnology Information (NCBI, http://www.ncbi.nih.gov) in the United States, which is considered to be the central database of genomic information. Several other similar databases are spread across Europe and Japan, but all of the core databases exchange data at intervals of 24 hours with the NCBI. The *GenBank* (http://www.ncbi.nih.gov /genbank) is the main NCBI database that stores all publicly available DNA sequences (from small sequences to entire genomes), RNA, and proteins. Apart from *GenBank*, which collects all incoming sequences, other databases at the NCBI have the information organized in different ways. For example, the *UniGene* (http://www.ncbi.nih.gov/unigene) groups together all the partial sequences of the transcriptome of an organism into clusters, where each cluster is the consensus sequence of a gene, whereas in *GEO* (http://www.ncbi.nih.gov/geo) it is possible to analyze the expression of a given gene in all public microarray data. Also at the NCBI, the *Gene* database (http://www.ncbi.nih.gov/gene) comprises only the reference sequences, or the most representative sequence of a transcript, edited, and inspected by a curator and anchored to the genome. It is often the best database to use in order to avoid redundancy in a universe where there is so much information available. Other banks are specific to an organism, such as the *Online Mendelian Inheritance in Man* (OMIM, http://www.ncbi.nih.gov/omim), which was created to catalog all the genes and alleles related to diseases and other human characteristics, as well as providing a detailed technical and bibliography for each feature. The existence of these databases, known as secondary databases, is as important as preserving the original data in *GenBank*.

Databases that assemble metabolic pathways are also available, such as *Kegg* (http://www.genome.jp/kegg), a database provided by GenomeNet Japan (http://www.genomenet.jp). This database couples metabolic pathways with information about the organisms in which they occur, or not! The consequences of the absence of pathways, such as the biosynthesis of essential amino acids (Guedes *et al.*, 2011), are likely to be investigated only with databases of complete genomes that are now available.

Several tools developed by bioinformatics allow access and analysis of the databases. The most popular tool for the comparison of DNA sequences with the database of sequences is *BLAST* (http://www.ncbi.nih.gov/blast) or *Basic Local Alignment Search Tool* (Altschul *et al.* 1990). By means of this algorithm we can compare a sequence of DNA or a protein under investigation (*Query*) with all the sequences in the public domain. It is important to note that the program *BLAST* does not aim to conduct a comparison along the total length of the compared molecules, but only to identify in the database the presence of a partial sequence that is sufficiently similar to that searched. Therefore it rapidly discards the unproductive results and extends the neighborhood of the region of similarity detected as far as possible. The result of that search returns, amongst the deposited sequences (DNA, RNA or proteins), those with higher scores on measurements of local similarity. Thus, several regions of DNA can be annotated by *BLAST*, this result being used to suggest, or assign a function to any segment of DNA, due to the fact that the observed similarity was too high as compared with what would be expected by chance. It is interesting to note that if we used one dinucleotide, "AT," for example, to search the

Genbank sequences, the expected number of targets would be very high, since it is expected that the dinucleotide will be found at random chance in a vast number of deposited sequences. If our query sequence was more complex, for example, 144 bases, the chance of finding at random another identical sub-sequence of 140 bases would be infinitely small. The E-value, a parameter calculated by *BLAST*, expresses this difficulty, the lower its value, the lower the chance of such a comparison being found by pure coincidence. In queries that return slightly different sequences, but with too small an E-value to be explained by chance, an alternative hypothesis is assumed: that the sequences have a common origin and then diverged slightly throughout evolution, which may or may not account for functional modifications.

There are several *BLAST* methods (Figure 9.5). The most enquiring and of great importance in gene discovery is that where both the *Query* sequence and the database members (*Subject*) are sequences of nucleotides, but comparisons are made between amino acids encoded by these sequences. In this program, before checking for similarity, the six possible translations are made for each nucleotide sequence, thus both the query sequence and all those found in the database are transformed into six proteins (translation initiating at the bases 1, 2 or 3 of each strand, the "+" and the "−" strands). This method, called *tBLASTx*, allows the pair "*Query* protein–*Subject* protein" to be returned and is very powerful, because the proteins of two organisms are generally

---

telomerase reverse transcriptase isoform 1 [Homo sapiens]
Sequence ID: ref|NP_937983.2| Length: 1132 Number of Matches: 1
▷ See 6 more title(s)

Range 1: 405 to 940 GenPept Graphics   ▽ Next Match △ Previous Match

Score	Expect	Method	Identities	Positives	Gaps	Frame
139 bits(350)	2e-30	Compositional matrix adjust.	125/564(22%)	233/564(41%)	43/564(7%)	+2

```
Query  1025  FNYYLTKSCPL-----PENWRERKQKIENLINKTREEKS--KYYEELFSYTTDNKCVTQF  1183
             +    L   CPL     P    ++K +  +   EE +  +  +L   +   V F
Sbjct  405   YGVLLKTHCPLRAAVTPAAGVCAREKPQGSVAAPEEEDTDPRRLVQLLRQHSSPWQVYGF  464

Query  1184  INEFFYNILPKDFLTGR-NRKNFQKKVKKYVELNKHELIHKNLLLEKINTREISWMQVET  1360
             +      ++P     R  N + F +  KK++ L KH  +    L  K++ R+ +W++
Sbjct  465   VRACLRRLVPPGLWGSRHNERRFLRNTKKFISLGKHAKLSLQELTWKMSVRDCAWLRRSP  524

Query  1361  SAKHFYYFDHE-NIYVLWKLLRWIFEDLVVSLIR•FFYVIEQQKSYSKTYYYRKNIWDVI  1537
                +H      +L K L W+    VV L+R FFYVIE     ++ ++YRK++W  +
Sbjct  525   GVGCVPAAEHRLREEILAKFLHWLMSVYVVELLRSFFYVIETTFQKNRLFFYRKSVWSKL  584

Query  1538  MKMSI-ADLKKETLAevqekeveewkkS-LGFAPGKLRLIPKKTTFRPIMTFNKKIVNSD  1711
             + I  LK+  L E+ E EV +  +++       +LR IPK   RPI+  +V +
Sbjct  585   QSIGIRQHLKRVQLRELSEAEVRQHREARPALLTSRLRFIPKPDGLRPIVNMD-YVVGAR  643
```

Figure 9.5   Results of a similarity search with the *BLAST* program. The first cloning of a telomerase catalytic subunit was from the protozoan *Euplotes*. Interestingly, the nucleotide similarity search (*BLASTn*) between the ciliated and human genes does not find sufficient similarity in any target sequence. However, the *BLASTx* method performs the translation of the nucleotide sequence from *Euplotes* (*Query*) in six possible amino acid sequences (three from the "+" strand and three from the "−" strand). The figure shows that the second phase of possible translation (*Frame* +2) aligns with human telomerase (*Subject* of the search or *Subjct*). The number of alignments expected by chance is low, 2e^{-30}. Owing to local alignment searches such as this, the discovery of human telomerase was made in 1997.

more similar to each other than the nucleotide sequences encoded by them. In this analysis, only one of the six translations is of biological significance, the others generate results that are disregarded. The *tBLASTx* method has been used in gene discovery numerous times, for example for identification by similarity of the catalytic subunit of human telomerase (Figure 9.5) as soon as this enzyme had been cloned from the protozoan *Euplotes* (Meyerson *et al.*, 1997). Other *BLAST* methods seek homology between nucleotide sequences (*BLASTn*), protein sequences (*BLASTp*), or between nucleotide sequences (which are internally translated by the program) versus proteins (*BLASTx*). Another variety of *BLAST* is the *PSI-BLAST*, which in a first round finds the most similar proteins to the *query* and goes on to identify conserved regions among the top search results; in subsequent searches, it masks the regions not conserved in *query* and performs the next search taking into account only the conserved regions. For the subsequent alignments, it does not rely on a general scoring system (BLOSUM64 matrix), but on the amino acid frequency observed in the top search results (PSSM matrix).

In databases, there is also a wide variety of information about molecular structures, differential gene expression, genetic diversity, evolution, and so on, that can be extracted by bioinformatics. A major challenge is the development of procedures by which such data can be "inserted" and "extracted" in/from secondary databases, by researchers. There are various tools that are available in the NCBI and other centers, but there is much scope for the development of specific procedures. Tools newly developed include databases of genes classified according to their evolutionary history (*COG* – NCBI, *Kegg*), algorithms for comparison of whole genomes (*ACT* – *Artemis Comparison Tool*), tools for finding structural similarity of proteins, regardless of the primary sequence (*VAST* – NCBI), and so on.

As the sequencing of genomes of many species is being accomplished, comparative genomics becomes increasingly more important and computational procedures for correlation between organisms at the molecular level become essential. Comparative research has been used for functional studies of the genome, for example, differential analysis of genes of pathogenic and non-pathogenic *Escherichia coli* (Perna *et al.*, 2001) allowed the identification of those genes related to the cause of bacterial disease (Jimenez-Sanchez *et al.*, 2001). Other studies allowed the identification of DNA sequences and functional elements responsible for significant differences among species, such as between man and chimpanzee (Ebersberger *et al.*, 2002). Comparative genomics has demonstrated that, in the evolutionary history of prokaryotes, several DNA segments were exchanged between different species, a process named horizontal transfer. Other applications of the comparative analysis of genomes are emerging: development of tissues and organs, the basis of resistance to infectious diseases, cancer prognosis, and so on. For each of these purposes, new bioinformatics tools are built and many of them are available via servers on the Internet.

A new discipline derived from genomics, pharmacogenomics, has already received significant investment in various companies in order to develop new drugs from genomic analysis. Much of the research in pharmacogenomics depends on the identification of inter-individual variations in humans for localization of genes related to disease susceptibility or resistance to drugs. Some companies have private databases containing these genetic variants, mostly SNPs (Single Nucleotide Polymorphisms), which correspond to differences in one nucleotide position. The NCBI has a public database of SNPs in different organisms, and in the human species there are more than 4 million SNPs cataloged. Other research groups and enterprises have invested heavily in the identification of SNPs in model organisms such as the mouse, specifically for pharmacogenomics applications. To collect SNPs, new studies can be pursed using molecular biology methods and bioinformatics tools, searching for association between alleles and distinct characteristics that are important for the development of new personalized medicines and more precise treatments without side effects.

The current organization of knowledge in secondary databases is very interesting because, even before an organism's genome is completely sequenced, much of the analysis of the molecular functions and biological processes present can be readily obtained by comparison. An important initiative was the creation of specific terms by the *Gene Ontology* consortium (http://www.geneontology.org) (Figure 9.6). The *GO terms* (pronounced like the verb "to go") have hierarchical relationships in the form of a tree, where the leaves specify either the functions or general processes. In parallel with the progress of construction of ontologies, another consortium, *GOA (Gene Ontology Annotation)*, assigns *GO terms* to sequences. New approaches have appeared such as *BLAST2GO* (Conesa *et al.*, 2005), which classifies populations of sequences, such as those from different transcriptomes, according to the occurrences of *GO terms*, and can depict the enrichment of transcripts in relation to certain functions or processes, in response to a challenge with a drug, for example.

Annotation of a Complete Genome

The use of *BLAST* software, as explained earlier, concerns the identification and characterization of nucleotide and protein sequences in bioinformatics. However, some proteins are unique to certain taxonomic groups, or have not yet been characterized. A very simple way to identify a putative protein coding region is the unexpected absence of the triplets TAA, TAG, and TGA in DNA, because these are

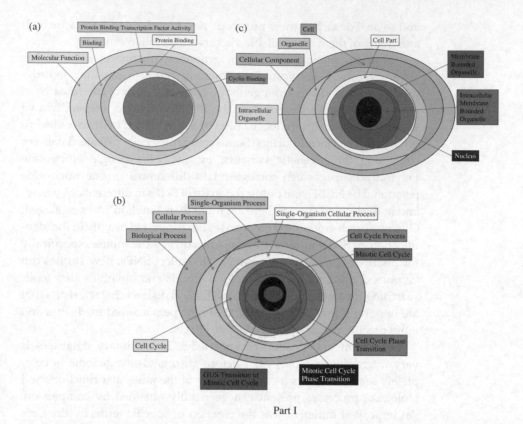

Part I

Figure 9.6 Gene Ontology. The hierarchy of *GO terms* related to (a) molecular function, (b) biological process, and (c) cellular component, associated with a protein that controls cellular proliferation, *cdk1*. Associated *GO terms* reveal that *cdk1* functions through cyclin binding, plays a role in G1/S transition of the mitotic cell cycle and occurs in the nucleus. It is easy to recover information associated to all children of a given *GO term*, for example, cell cycle (GO:0007049) (http://www.ebi.ac.uk/QuickGO/). (See color figure in color plate section).

transcribed into UAA, UAG, and UGA, the stop codons. Their absence in significantly long stretches is a good indication of the encoding of proteins in that particular stretch of DNA. Other parameters such as frequency of some dinucleotides and characteristics that are sometimes extracted by artificial intelligence suggest the encoding of proteins in the DNA, which are then termed "predicted" proteins by the software. Interestingly, although no one has studied them in detail, these proteins can be found in various organisms, which is another piece of evidence for their functional existence, and at this point they become referred to as hypothetical proteins. Thus, when an operator uses a software for annotation of genomes such as *Artemis* (Figure 9.7), he/she focuses the attention on sections without stop codons and, with the aid of gene prediction software (*Glimmer*, for instance) and software for alignment with sequences of other organisms, such as *BLAST*, the operator can identify

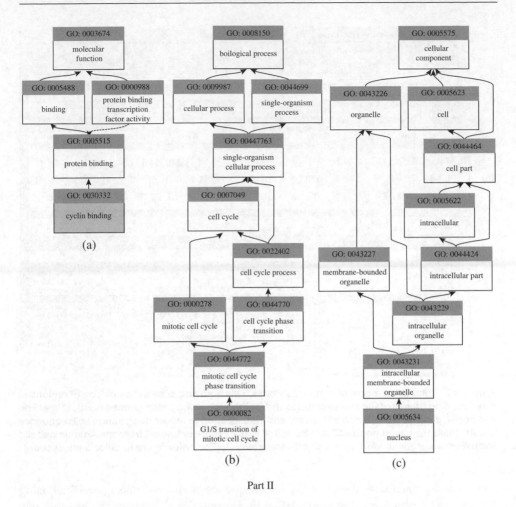

Part II

Figure 9.6   (continued)

protein coding regions, whether of known or hypothetical functions. The annotation of a genome with introns is a bit more complex, since it adds the difficulty of correctly determining the gene model, that is, the regions where exons and introns are located, and often there are several isoforms of RNA processing (Figure 9.4).

Computational System with Chained Tasks Manager (*Workflow*)

Recently, it has become possible to integrate bioinformatics tasks with a computational system manager, coupling chained tasks. This system not only coordinates tasks, but also facilitates the use of several concatenated software programs, integrating the formats of data and results. Thus, it

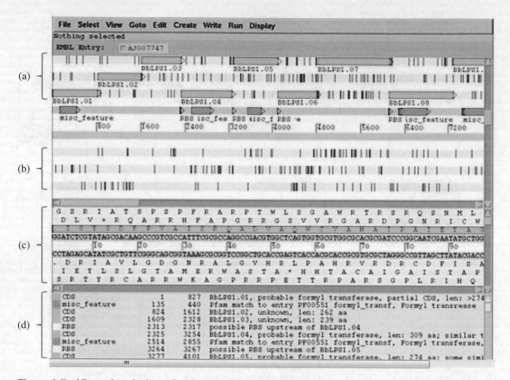

Figure 9.7   View of a window of software *Artemis*. This software allows the marking of regions in three reading frames of the two strands (a) and (b) in which nearby stop codons (vertical lines) do not occur, putative genetic regions (cyan) and visualization of the working amino acid sequence (c). Annotations based on *BLAST* can be edited in (d). Figure extracted from the *Artemis* website (http://www.sanger.ac.uk/resources/software/artemis). (See color figure in color plate section).

facilitates the execution of a sequence of chained tasks (*workflow*) using programs that are installed in computational *clusters*. The most widely known and frequently used online system with several workflows for bioinformatics centers is *Galaxy* (Schatz, 2010). It was initially developed for genomics, but it is now used for various applications. The simplicity of its use lies in the implementation of graphical environments viewed in a web browser, comprising control panels for submitting tasks, choosing the desired sequence of software (*pipeline*), and helping project management. The results obtained by various analyses are all integrated, making reference to each other. In addition, and more importantly, the installation of new software in the system is very user friendly, integrating it with the others automatically.

Applications for Studies in Plants

Data presented in Figure 9.8 show the large amount of information within the kingdom Viridiplantae (green plants). There are millions of

Database name	Entries
Nucleotides	16,713,333
ESTs	24,893,532
Proteins	2,836,727
Structures	3,108
Genomes	938
SNPs	37,270,327
Domains	1,551
Microarrays	67,714
Genes in Microarrays	3,543,347
Genes	795,035
Homologues	12,488
NGS Experiments	19,212
Taxonomic groups	145,770

Figure 9.8   Entries in databases related to organisms of the kingdom Viridiplantae (green plants). Query made to NCBI Taxonomy in October 2013 (http://www.ncbi.nih.gov/taxonomy).

known entries of nucleotide and protein sequences, and thousands of proteins with the 3D structure determined. Thousands of microarray experiments that have been deposited in public databases such as *GEO* (Barrett *et al.*, 2013) can be analyzed. More than half a million edited entries without duplication are found in the *Gene* database. There is information deposited on sequences of nearly 150 000 organisms. The knowledge accumulated so far greatly facilitates data analysis in projects with new organisms or with environmental samples. This accelerates more and more the system analysis of the data, in which bioinformatics plays a strategic role.

Final Considerations

The 19th century was the era of a revolution in biology after publication of the theory of evolution, which made it possible to investigate and understand the origins and functioning of organisms. Since the late 20th century, the biological knowledge that has emerged has led to major advances in genomic approaches and their derivatives, resulting in the generation of biological data on a large scale. In the 21st century, we can contemplate the advances of bioinformatics supporting the analyses and storage of all this information arising from the "omics." Bioinformatics depends on the efforts of the human intellect to develop algorithms and routines that allow the use of the logic capacity of computers to process and display the "omics" megadata. However, there is currently a high demand for expansion of this logical treatment of the data, coupled with the possibility of scientists establishing the present systemic hypotheses, which places bioinformatics in a central position in the exploitation of the several "omics." There is still much to do in terms of software, but increasingly they are being directed to test hypotheses and to develop models to find out what is behind the most complex biological processes.

References

Altschul, S.F.; Gish, W.; Miller, W.; *et al.* 1990. Basic local alignment search tool. Journal of Molecular Biology, 215 (3):403–10.

Barrett, T.; Wilhite, S.E.; Ledoux, P.; *et al.* 2013. NCBI GEO: archive for functional genomics data sets – update. Nucleic Acids Research, 41 (D1): D991–D995. Published online 26 November 2012; doi: 10.1093 /nar/gks1193 PMCID: PMC3531084.

Conesa, A.; Götz, S.; García-Gómez, J.M.; *et al.* 2005. Blast2GO: a universal tool for annotation, visualization and analysis in functional genomics research. Bioinformatics, 21 (18): 3674–3676. Epub 2005 Aug 4. PubMed PMID: 16081474.

Ebersberger, I.; Metzler, D.; Schwarz, C.; Pääbo, S. 2002. Genome-wide comparison of DNA sequences between humans and chimpanzees. American Journal of Human Genetics, 70 (6):1490–1497. Epub 2002 Apr 30. PubMed PMID: 11992255; PubMed Central PMCID: PMC379137.

Guedes, R.L.; Prosdocimi, F.; Fernandes, G.R.; *et al.* 2011. Amino acids biosynthesis and nitrogen assimilation pathways: a great genomic deletion during eukaryotes evolution. BMC Genomics. 12 (Suppl 4): S2. Epub 2011 Dec 22. PubMed PMID: 22369087; PubMed Central PMCID: PMC3287585.

Jimenez-Sanchez, G.; Childs, B.; Valle, D. 2001. Human disease genes. Nature. 409 (6822): 853–855. PubMed PMID: 11237009.

Meyerson, M.; Counter, C.M.; Eaton, E.N.; *et al.* 1997. hEST2, the putative human telomerase catalytic subunit gene, is up-regulated in tumor cells and during immortalization. Cell, 90(4): 785–795; PubMed PMID: 9288757.

Perna, N.T.; Plunkett, G., III; Burland, V.; *et al.* 2001. Genome sequence of enterohemorrhagic *Escherichia coli* O157:H7. Nature, 409 (6819): 529–533; erratum, Nature, 410 (6825): 240; PubMed PMID: 11206551.

Prosdocimi, F.; Santos, F.R. 2004. Sobre bioinformática, genoma e ciência. Ciência Hoje, 35 (209): 54–57.

Schatz, M.C. 2010. The missing graphical user interface for genomics. Genome Biology, 11 (8): 128; doi: 10.1186/gb-2010-11-8-128. Epub 2010 Aug 25. PubMed PMID: 20804568; PubMed Central PMCID: PMC2945776. https://usegalaxy.org/ (accessed 3 February 2014).

Soneson, C.; Delorenzi, M. 2013. A comparison of methods for differential expression analysis of RNA-seq data. BMC Bioinformatics. 14: 91; doi: 10.1186/1471-2105-14-91. PubMed PMID: 23497356; PubMed Central PMCID: PMC3608160.

Trapnell, C.; Roberts, A.; Goff, L.; *et al*. 2012. Differential gene and transcript expression analysis of RNA-seq experiments with TopHat and Cufflinks. Nature Protocols, 7 (3): 562–578; doi: 10.1038/nprot.2012.016. PubMed PMID: 22383036; PubMed Central PMCID: PMC3334321.

10 Precision Genetic Engineering

Thiago J. Nakayama,[a] Aluízio Borém,[a] Lucimara Chiari,[b]
Hugo Bruno Correa Molinari,[c] and Alexandre Lima
Nepomuceno[d]

[a]Department of Crop Science, Federal University of Viçosa, Viçosa,
MG, Brazil
[b]Embrapa Beef Cattle, Campo Grande, MS, Brazil
[c]Laboratory of Genetics and Biotechnology, Embrapa Agroenergy,
Brasília, DF, Brazil
[d]Embrapa Soybean, Londrina, PR, Brazil

Introduction

The barrier established by interspecific reproductive isolations has
always limited the available genetic base for plant breeding. Only the
natural allelic variations within the gene pool and the random variations
induced by radiation or chemical mutagenesis have been available to
plant breeders for the development of more productive and commercial
varieties adapted for different environments.

However, rapid advances in the field of genetics have enabled the
sequencing of entire genomes within days/hours/minutes, permitting
the exploration of the entire genetic diversity of a determined species
and the identification of new metabolic routes and genes in species
never before studied for agricultural, medicinal, and industrial use.

The first plant sequenced (*Arabidopsis thaliana*) required approximately 10 years before the first rough draft of its genome was
presented. Today, with the use of next-generation sequencing (NGS)
(e.g., Oxford Nanopore, PacBio RS, Ion Torrent, and Ion Proton, etc.)
of DNA and powerful bioinformatics and computational modeling
programs, genomes can be sequenced, annotated and related to
specific phenotypic traits of a particular genotype in a few weeks.
These advantages, along with great reductions in sequencing costs,
have facilitated the generation of a growing volume of data and have
enabled the thorough study of genomes, transcriptomes, proteomes,
and metabolomes and their inter-relations, which are responsible for

Omics in Plant Breeding, First Edition. Edited by Aluízio Borém and Roberto Fritsche-Neto.
© 2014 John Wiley & Sons, Inc. Published 2014 by John Wiley & Sons, Inc.

the diversity of phenotypic responses in vegetable and animal species and microorganisms.

The genetic information generated by high-performance sequencing technologies has provided the foundation for the development of new Genetic Engineering (GE) strategies aimed at increasing yield and tolerance to biotic and abiotic stresses in plants. However, until recently, the available GE tools merely permitted the modification of larger DNA sequence blocks that could only be randomly inserted within the species genome.

Recent advancements in GE have allowed the construction of new species variations from site-directed modifications, including specific mutations, insertions, and substitutions of genes and/or gene blocks. These variations are a powerful alternative for the development of commercial varieties with new traits of high agronomic value and/or high added value for industry or medicine.

Precision Genetic Engineering (PGE) focuses on the development of site-directed modification methods in specific DNA sequences to introduce new phenotypic traits in the individuals to be genetically modified. The use of PGE in plants is fairly recent and has been particularly focused on aggregating new traits in commercial varieties, which was impossible to achieve via traditional breeding or with the necessary "surgical precision" by classical plant transformation techniques.

Modifications in specific DNA sequences begin with the generation of a DNA break (DNA double-stranded break, DSB). Genetically modified nucleases have been engineered to identify a specific sequence of the target genome and to catalyze DSB, facilitating the incorporation of the desired DNA modifications at or close to a break.

Several strategies have been developed to access specific sites of the genome of a species and to generate DSBs to promote the desired genetic modification (Figure 10.1).

Mutagenesis is probably the easiest type of site-directed modification. To begin mutagenesis, a nuclease is engineered to produce a DSB at a specific chromosomal site. Mutagenesis occurs by simply expressing the nuclease in the vegetable cell. The DSBs created by the nuclease are repaired by non-homologous end joining (NHEJ), which very often results in small indels at the DSB. The introduction of indels in the target site in an open reading frame (ORF) of a gene, particularly inside an exon close to the 5' end, can cause reading frame changes, creating a non-functional knockout gene. Molecular analysis can then be carried out on populations of transformed vegetable cells to detect the desired changes.

The substitution or repair of DNA sequences is the most powerful form of PGE due to the possibility of altering a gene function instead of simply knocking them out. The use of homologous recombination (HR) to repair nuclease-mediated DSBs enables specific alterations in a gene

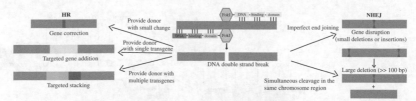

Figure 10.1   Modifications in specific DNA sequences are initiated with generation of a DNA double-stranded break (DSB). The diagram shows the possible results of a nuclease-mediated DSB. The presence of a sequence donor can stimulate the homologous recombination (HR) pathway, allowing donor sequence integration (on the left). DSBs also can be repaired by non-homologous end joining (NHEJ) (on the right), which can result in small insertions or deletions (indels) at the DNA break. These indels can interrupt the reading frame of the target gene, knocking it out. Two simultaneous DSBs made on the same chromosome can lead to a large deletion. (See color figure in color plate section).

sequence. In this case, the DSB is repaired using a DNA donor containing sequences that are homologous to those flanking the site of the break. This strategy is more complex than mutagenesis because the molecules of the donor DNA must be available in the cell at the moment that the nucleases mediate cleavage. The HR pathway uses the homologous region of the donor end to repair the DSB, thus incorporating the donor sequence in the target chromosome/gene. The donor DNA sequence can be altered to obtain the desired effect, including changes in promoter regions or modifications in the sequence that alter the catalytic activity of the encoded protein.

For gene insertion, one or more transgenes are introduced in a specific chromosomal site by means of HR. The gene insertion sites are usually directed to genomic regions that favor high expression levels. In addition, the insertion of multiple transgenes in the same chromosomal locus subsequently facilitates the block transfer of all transgenes to other varieties as they behave as one locus.

Site-directed structural alterations (e.g., deletions and inversions) are also of interest. In this case, one or two different nucleases may cleave adjacent sites along the chromosome. Some of these events can be repaired by fusion of the respective breaking points, thereby deleting/excluding the sequence that separates the two cleavage points. This method can be useful to remove pools of undesirable genes for important traits such as productivity or tolerance to abiotic stresses. Alternatively, the fragment to be released can be inverted between the two cleavage sites before DSB repair occurs. Similar methods can be used to stimulate chromosomal translocations if the two cleavage sites are on different chromosomes.

To modify genes in plants using PGE, four stages are necessary: (i) the design and development of a construct containing a genetically

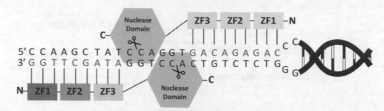

Figure 10.2   Scheme illustrating three ZFN modules, which recognize 9 bp upstream and 9 bp downstream of the region where the DSB is desired. Each ZFN module is linked to the catalytic domain of a nuclease that will catalyze the DSB. (See color figure in color plate section).

modified nuclease to recognize the target region and to generate the DSB; (ii) the transfer of the construct containing the donor molecule to the plant (typically via genetic transformation); (iii) induction of nuclease expression; and (iv) the selection of plants with alterations in the targeted DNA sequence. The genetically modified nuclease used for site-directed modification can be eliminated later leaving no traces of its construction. The cassette for the nuclease and guide elements (protein for TALEN/ZFN or RNA for CRISPRs) expression is inserted in a different locus from the target gene.

Heterozygous T0 mutant will be generated and T1 segregating plants than can be genotyped to identify individuals that are homozygous for the mutation that do not carry the nuclease/guide elements sequences.

In recent years, the greatest advancements in PGE have focused on the development of proteins that specifically identify and promote DSBs in loci of interest. To achieve this critical step, three types of enzymes have been subjected to genetic modification/engineering strategies: zinc finger nucleases (ZFNs) (Figure 10.2), transcription activator-like effector nucleases (TALENs) (Figure 10.3) and meganucleases, also known as LAGLIDADG homing endonucleases (LHEs) (Figure 10.4). The three strategies follow the same general principle; all three proteins consist of a DNA binding domain, which is responsible for target site specificity, and a domain with endonuclease activity, which functions like a restriction enzyme to produce the DSB.

Zinc Finger Nucleases (ZFNs)

Technologically ZFN began with the pioneering work done by researchers from the Johns Hopkins University when they were trying to produce new restriction enzymes (Kim, Cha, and Chandrasegaran, 1996). Their studies began around the middle of the 1990s and were focused on type IIS restriction enzymes such as *Fok*I, which recognize specific DNA sequences and cleave several base pairs 5′–3′ downstream

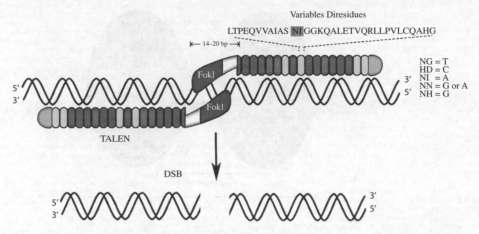

Figure 10.3 Scheme illustrating TALENs functional motifs. TALENs have 16–20 highly conserved repetitive peptidic modules, each of which is composed of 33 or 34 amino acids. The repeated modules differ at positions 12 and 13 (diresidues), which specifically recognize one of four nitrogenated bases. (See color figure in color plate section).

of the recognition site. The goal of the research was to fuse the *Fok*I catalytic domain to a protein domain that recognizes and binds to DNA to alter restriction enzyme specificity. These researchers selected zinc finger (ZF) proteins as the protein domain to recognize DNA. ZF proteins normally recognize blocks of three nitrogenated bases in single sequential order. The ZF protein units can be combined in specific arrangements, typically with 3–6 ZFs, thus enabling the recognition of single DNA sequences of 9–18 bp and targeting the creation of DSBs in the recognition region.

Two classic studies demonstrated the usefulness of ZFNs for specific alterations in the genomes of *Drosophila melanogaster* and humans (*Homo sapiens sapiens*): Bibikova *et al.* (2003) and Porteus and Baltimore (2003). Genetic alterations mediated by this DSB generation strategy have also been efficient in other animal models, including rats (*Rattus norvegicus*), mice (*Mus musculus*), and zebrafish (*Danio rerio*) (Cui *et al.*, 2011; Geurts *et al.*, 2009; Mashimo *et al.*, 2010). The first positive results in plants were reported in *A. thaliana* and tobacco (*Nicotiana tabacum*) by Lloyd *et al.* (2005) and Wright *et al.* (2005), respectively.

Countless platforms can be used to build customized ZFNs. One of the first available public platforms was modular assembly (Wright *et al.*, 2006). This method involves the creation of arrangements containing multiple ZFs based on ZFs for which the DNA sequence that it binds is already known. Modular assembly can be used to easily construct ZFNs at low cost using standard subcloning techniques. However, many of the ZFNs created/engineered by this method do not have high activities (Ramirez *et al.*, 2008) because modular assembly treats each ZF as an

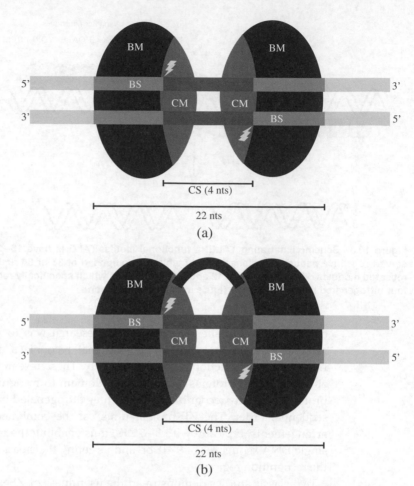

Figure 10.4 **Meganucleases may have one (a) or two (b) copies of the motif LAGLIDADG, which is formed by the binding module (BM) and the cleavage module (CM). The BM recognizes the DNA binding site (BS). The CM acts at the DNA cleavage site (CS), creating two 3′-adhesive ends of four nucleotides each. Meganucleases with a single motif are active as homodimers, while versions with two motifs are active as monomers. (Source: Hafez and Hausner, 2012). (See color figure in color plate section).**

independent unit without considering the influence of neighboring ZFs on DNA binding. To overcome this deficiency, platforms have been developed to select the best combinations by considering the individual and overall effects on final DNA binding characteristics.

The Oligomerized Pool Engineering Protocol (OPEN) has free access and selects ZFs based on a platform that identifies highly functional multiple ZF matrixes (Maeder *et al.*, 2008; Zhang *et al.*, 2010). The advantage of building ZFNs using the OPEN program is that combinations of ZF units that will work well as a group can be identified. The target sites of putative ZFNs can be identified at every 200 bp using OPEN.

However, the construction/engineering of ZFNs requires a highly qualified person and approximately 2–3 months of work. To overcome this limitation, a third platform has combined the ease of modular assembly with the selection of ZFs based on the effects of individual ZFs on assembly and vice versa. This platform, context-dependent assembly (CoDa), allows ZFNs to be engineered within 1 week. On average, CoDa permits the identification of one target site every 500 bp. Studies in zebrafish, *Arabidopsis* and soybean (*Glycine max* (L.) Merr.) have demonstrated that CoDa-derived ZFNs work effectively at approximately 50% of target sites (Curtin *et al.*, 2011; Sander *et al.*, 2011b).

Transcription Activator-like Effector Nucleases (TALENs)

Phytopathogenic bacteria of the genus *Xanthomonas* naturally contaminate a wide variety of species, including rice (*Oryza sativa*), citrus (*Citrus* spp.), tomato (*Solanum lycopersicum*), and soybean (*Glycine max*) (Boch and Bonas, 2010; Kay and Bonas, 2009). During infection, the *Xanthomonas* inoculated in vegetable cells produce a series of proteins known as transcription activator-like effectors (TALEs) (Boch and Bonas, 2010; Bogdanove *et al.*, 2010). TALEs modify the transcriptome of the host by binding to specific sequences of promoter regions, thus mimicking host transcription factors (Kay and Bonas, 2009). TALEs have a DNA binding domain that typically consists of 16–20 monomeric repetitions. Each monomer contains 34 amino acids and is highly conserved, with the exception of the hypervariable amino acid residues at positions 12 and 13, which are known as repeat-variable di-residues (RVDs). Recent computational and molecular biology analyses have permitted the decryption of the TALE code for DNA recognition (Boch *et al.*, 2009; Moscou and Bogdanove, 2009). Each RVD recognizes a different DNA base; for example, repetitions with the RVDs NI, HD, NG, or NN bind to adenosine (A), cytosine (C), thymine (T), and guanine (G) or adenosine (A), respectively.

The elucidation of the mechanism of DNA recognition by TALEs immediately attracted attention for its potential biotechnological applications (Bogdanove *et al.*, 2010). One of the initial experiments was to fuse the TALE binding domain to the *Fok*I endonuclease catalytic domain, thus creating TALENs. The fusion of the binding domains for native or personalized DNA sequences with *Fok*I enabled the production of specific DSBs (Christian *et al.*, 2010).

TALENs have been used to produce site-specific alterations in different species including *A. thaliana* (Cermak *et al.*, 2011), tobacco (Mahfouz *et al.*, 2011), rice (Li *et al.*, 2012), yeast (*Saccharomyces cerevisiae*) (Li *et al.*, 2011), zebrafish (Huang *et al.*, 2011; Sander *et al.*, 2011a), mice (Tesson *et al.*, 2011), and humans (Miller *et al.*, 2011; Maeder *et al.*, 2013). In addition to TALENs, TALEs have also been genetically modified to either activate or repress the expression of endogenous plant genes (Li *et al.*, 2012; Mahfouz *et al.*, 2012; Morbitzer *et al.*, 2010).

Several platforms are already publicly available for TALE engineering, including the *Golden Gate* assembly method, which enables the orderly gathering of sequences that code for TALE repetitions in a single reaction (Cermak *et al.*, 2011; Engler *et al.*, 2009; Sanjana *et al.*, 2012; Zhang *et al.*, 2011; Reyon *et al.*, 2012; Sander *et al.*, 2011a). Also, the development of a rapid and straightforward approach for the construction of designer TALE (dTALE) activators and nucleases with user-selected DNA target specificity has been reported. This platform has a set of plasmids that enable researchers to assemble repeat domains for any 14-nucleotide target sequence in one sequential restriction-ligation cloning step and in only 24 h. Moreover, a web tool was developed, called idTALE, to facilitate the design of dTALENs and the identification of their genomic targets and potential off-targets in the genomes of several model species (Li *et al.*, 2012). These platforms permit the rapid production of TALENs. One of the main advantages of TALENs is that a DNA target locus can be investigated at every 10 bp, allowing more extensive analysis than the flexibility permitted by ZFNs. Furthermore, most engineered TALENs are functional, making them preferable for many PGE applications.

Zhang *et al.* (2013) reported the high efficiency of TALENs for site-directed mutagenesis of plant genomes. These workers used protoplasts of tobacco (*Nicotiana tabacum*) and engineered TALENs to target the duplicate genes of acetolactate synthase (ALS; *SurA* and *SurB*). Specific substitutions of amino acids in ALS provide resistance to herbicides of the sulfonylurea and imidazolinone classes in a dominant manner (Tranel and Wright, 2002). The TALENs introduced specific mutations in ALS in 30% of transformed cells, and the insertion frequency in the target gene was approximately 14%. The efficiency of the method allowed the recovery of modifications without the necessity of selection or enrichment: 32% of the regenerated calluses had TALEN-induced mutations in ALS, and of the 16 calluses characterized in detail, all had mutations in an allele of each one of the duplicate ALS genes (*SurA* and *SurB*).

Meganucleases (LHEs: LAGLIDADG Homing Endonucleases)

LHEs differ from ZFNs and TALENs because the former enzymes naturally target genes and are encoded by mobile introns (Arnould *et al.*, 2011). LHEs form homodimers in which each identical subunit contains 160–200 amino acid residues. LHEs may also function as a single peptide formed by two monomers in tandem spaced by a third binding sequence (Stoddard, 2011). LHEs target DNA sequences of 20–30 bp, providing high specificity. Thus, LHEs have been developed as a platform for genomic modification. In contrast to ZFNs and TAL-ENs, the cleavage and DNA binding domains of LHEs are not clearly delimited. Attempts to engineer LHEs at their contact points with target DNA have been challenging and very often compromise endonuclease activity (Taylor *et al.*, 2012). Numerous methods have been developed to evaluate the DNA binding activity and specificity of LHEs (Gao *et al.*, 2010; Stoddard, 2011). These studies include high-performance tests in bacteria and yeast to evaluate cleavage and the subsequent reconstitution of reporter genes and computational approaches using several algorithms to evaluate how modifications in LHEs alter their DNA binding affinity. Owing to the greater engineering challenges present in this methodology, only a few academic groups and companies routinely customize LHEs to act at new target sites.

Clustered Regularly Interspaced Short Palindromic Repeats (CRISPR)

Not restricted to vertebrates as previously thought, prokaryotes also have adaptive immunity and the mechanism is RNA-based mediated by CRISPR/Cas systems (Clustered Regularly Interspaced Short Palindromic Repeats/CRISPR-associated proteins).

Basically, CRISPR/Cas systems are composed of *cas* genes organized in operons and long CRISPR arrays of identical repeats interspersed with unique spacer sequences derived from invading nucleic acids (protospacers). Cas proteins are involved in new spacer sequence acquisition (adaptation stage), CRISPR RNA biogenesis (biogenesis stage), and target interference (interference stage) (Wiedenheft, Sternberg, and Doudna, 2012), Cas9 has HNH and RuvC-like nuclease domains that cleave the complementary and noncomplementary target DNA strands, respectively (Jinek *et al.*, 2012). The CRISPR arrays codify precursor CRISPR RNAs (pre-crRNAs), which undergo maturation as CRISPR RNAs (crRNAs).

Of the CRISPR systems known, the Type II from *Streptococcus pyogenes* (Figure 10.5a) is one of the simplest, and has been the focus of new genome engineering technology. In this system, a partially complementary trans-activating crRNA (tracrRNA) and RNase III are needed to form a tracrRNA–crRNA complex to direct Cas9 to break the target DNA (Deltcheva *et al.*, 2011). In a simpler sgRNA/Cas9 system (Figure 10.5b), the duplex RNA structure requirement can be bypassed by using an artificial chimera small guide RNA (sgRNA) containing a designed hairpin (Linker loop) that mimics the tracrRNA–crRNA complex (Jinek *et al.*, 2012).

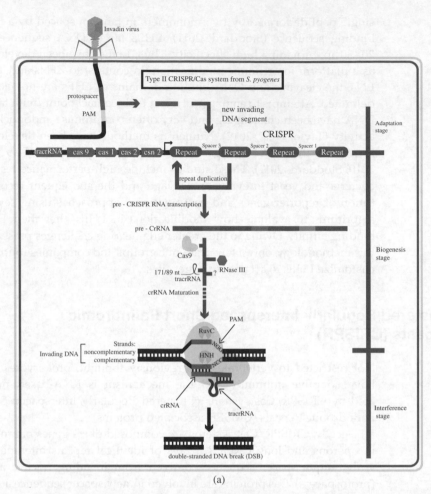

(a)

Figure 10.5 Type II CRISPR/Cas system pathway in *Streptococcus pyogenes* and targeted plant genome editing by type II sgRNA/Cas9 system. (a) Example of type II CRISPR/Cas system pathway in *S. pyogenes* in which an incorporation of a new invading DNA segment in the CRISPR locus occurs (adaptation stage); pre-CRISPR RNA transcription and crRNA maturation involving Cas9, RNase III, tracrRNA, and another possible unknown component (?) (biogenesis stage); and target DNA breaking by the tracrRNA–crRNA–Cas9 complex (interference stage). (b) Targeted plant genome editing by the type II sgRNA/Cas9 system involving coexpression of the cas9 host codon optimized protein bearing one nuclear localization signal (NLS) and gRNA expression by U6 polymerase III promoter. (See color figure in color plate section).

To make the sgRNA/Cas9 type II system from *S. pyogenes* effective in eukaryotic cells, as observed by Mali *et al.* (2013), Shan *et al.* (2013), and Li *et al.* (2013), some requirements are taken into account, such as: (i) Cas9 attachment with a nuclear localization signal (NLS) to perform Cas9 functionality in the nucleus; (ii) *cas9* host codon optimization for efficient heterologous expression; and (iii) sgRNA containing an approximately 20 bp spacer equal to the host cell noncomplementary target strand protospacer sequence.

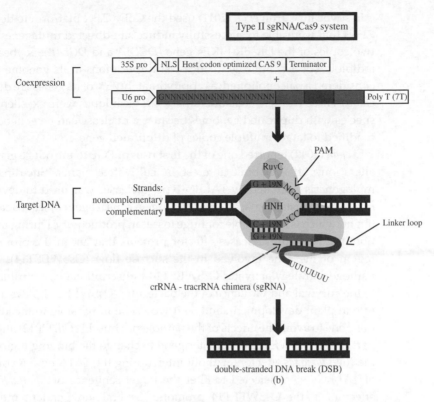

Type II sgRNA/Cas9 system

Figure 10.5   (continued)

For target DNA cleavage by the Cas9–sgRNA complex, a target proto-spacer with a short motif 5′-NGG-3′ (Protospacer Adjacent Motif (PAM), where N represents any nucleotide) adjacent on the 3′ side is needed (Jinek *et al.*, 2012) (Figure 10.5). In monocot and dicot plants, Cas9 and sgRNA expression cassettes have been generally driven by 35S and U6 promoters, respectively (Nekrasov *et al.*, 2013; Shan *et al.*, 2013; Li *et al.*, 2013; Feng *et al.*, 2013; Jiang *et al.*, 2013; Miao *et al.*, 2013). RNA polymerase III promoter (Wang *et al.*, 2008), for example U6, has been used due to the ability to drive hairpin sgRNA expression. Although the U6 promoter constrains the first position in the RNA transcript to be a "G," Jinek *et al.* (2012) demonstrated *in vitro* that few mismatches in the 5′ side of the protospacer can be tolerated. The system described here in principle allows any sequence of the form 5′-N(21)-GG-3′ to be targeted.

Implications and Perspectives of the use of PGE in Plant Breeding

There are several examples of the use of PGE in model plants; recently, the practical applications of its use in cultivated species have begun to appear.

In soybean, Curtin *et al.* (2011) used the CoDa ZFN platform to design ZFNs that were able to successfully induce site-directed mutagenesis in two copies of the DICER-LIKE4 gene (*DCL4A* and *DCL4B*). Soybean is a diploid species, but due to ancestral duplications in its genome, it is considered a paleopoliploid, as approximately 75% of its genes are duplicated. These workers concluded that PGE techniques were excellent for species with duplicated genomes because a nuclease can be genetically modified to target multiple copies of duplicated genes.

Li *et al.* (2012) reported one of the first uses of PGE to introduce genetic alterations of agronomic interest in cultivated plants. Site-directed mutagenesis mediated by TALEN-type nucleases was used to develop rice plants resistant to *Xanthomonas oryzae*, a causal agent of leaf bacterial blight, which is responsible for huge losses in productivity. During infection, the bacterium releases effector proteins that act in the promoter region of the gene involved in the sucrose flow (OsSWEET14). This pathogen strategy activates OsSWEET14 transcriptionally, contributing to the survival and virulence of the bacterium. OsSWEET14 plays a key role in plant development, and so it would be impossible to inactivate this gene to avoid the effects of the pathogen. Thus, Li *et al.* (2012) altered part of the OsSWEET14 promoter gene to change the binding region of the *Xanthomonas* effectors without interrupting the TATA box. A couple of TALENs were selected to alter the target sequence of *Xanthomonas* effectors in the OsSWEET14 promoter. Several independent mutant lines containing TALENs were obtained, some of which exhibited resistance to *X. oryzae*. The resistant rice plants were analyzed and were determined to be mutant homozygous or heterozygous at the target site. In addition, these plants were morphologically identical to the parental plants, indicating that the mutations did not result in adverse phenotypes typical of OsSWEET14 inactivation.

One of the potential advantages of PGE is the capacity to create an individual with a new genetic variation without introducing a transgene. In experiments involving ZFNs and TALENs, respectively, Curtin *et al.* (2011) and Li *et al.* (2012) genotyped T1 segregant plants to identify those that maintained their site-directed mutations and segregated the parts of the integrated genetically modified (GM) nucleases. However, to confirm the complete absence of fragments of nucleases in these lineages, complete sequencing of the genomes of the plants obtained is required.

Alternatively, virus-based methods permit the transient production of nucleases. Marton *et al.* (2010) demonstrated the transient release of ZFNs in tobacco and petunia using an expression system based on tobacco rattle virus (TRV). The TRV was able to move to the meristems, enabling the changes mediated by the ZFNs to be transmitted to the next generation. Because the host plant genome is not genetically modified, this virus-based delivery system may offer regulatory advantages for the commercialization of the resulting plants, in contrast to transgenic varieties.

Shukla *et al.* (2009) produced ZFNs to specifically modify the *IPK1* corn gene, which encodes the inositol-1,3,4,5,6-pentakisphosphate enzyme. This enzyme is an important catalyst in the phytate biosynthetic pathway and contributes to an increase in total phosphorus levels in corn seeds. The reduction of phytate content is of agricultural interest because phytate is considered an antinutritional factor because it interacts with minerals such as calcium, zinc, phosphorus, iron, and copper, along with some proteins, forming insoluble complexes and diminishes their bioavailability, in addition to being a pollutant in animal residues. Thus, to modify the activity of the *IPK1* gene, these workers evaluated a total of 66 ZFNs for five positions along the *IPK1* locus. Four ZFNs were selected to act on the second exon of this gene based on their capacity to produce DSBs and NHEJ repairs in corn cell cultures. The *IPK1* locus was cleaved via ZFNs and repaired via HR using fragments containing a herbicide-resistant gene flanked by *IPK1*-complementary sequences as donor sequences. Approximately 600 herbicide-resistant transformed calluses were examined, and several monoallelic insertions and one biallelic insertion from the herbicide resistance gene were observed in the *IPK1* locus. The descendants exhibited the expected segregation frequencies, indicating that the insertions were effectively transmitted to the next generation.

Structural alterations such as large deletions in target sequences in cultivated species have also been reported using PGE. These studies have included efforts to eliminate reporter genes and marker genes in GM plants. Petolino *et al.* (2010) demonstrated that the removal of a reporter gene (*GUS*) flanked by ZFN cleavage sites was achieved after crossing with a plant bearer of ZFN. The progeny presented complete deletions of the reporter gene. The implications of this technology for plant breeding include the capacity to remove undesired marker genes after integration of the transgene, in theory facilitating market liberation without regulatory restrictions. Another application is the deletion of large, highly repetitive DNA regions. Such site-directed deletions could become an important tool for the generation of new traits in plants cultivated by the elimination of repetitive DNA or other undesired loci.

The combination of multiple transgenes in a single genome, also known as gene stacking, is normally obtained using classic genetic improvement. Depending on the number of transgenic loci being combined, large progeny populations must be selected to obtain a plant with the desired loci combination. Gene stacking of binding transgenic loci is desired by breeders because it facilitates the introduction of new transgenic traits for the development of new commercial cultivars. PGE can simplify the insertion of these new transgenic traits in tandem with other, already present transgenes in the target genome, thus permitting gene stacking of transgenes with higher precision.

Corn varieties obtained using PGE technologies have already been commercially released. For instance, the corn LY038 from the Monsanto Company was approved by the USDA in 2006. This corn contains high lysine levels, which supplements corn-based rations used for pigs and birds. The *CORDAPA* gene, which is responsible for the increased lysine content, was introduced beside an antibiotic-resistance gene used to select transformed cells. Subsequently, using the Cre-Lox system and site-specific recombination technology, the antibiotic resistance gene was removed (Wang *et al.*, 2011). In the Cre-Lox system, the Cre protein, a site-specific recombinase, catalyzes DNA recombination between specific regions. In this case, the LoxP regions contain binding sites for the protein Cre, which encloses the central sequence in a directional manner, promoting recombination and excluding the sequence flanked by the LoxP sequences.

PGE enables a great variety of specific alterations in vegetable genomes, and due to its various applications, it is possible that the varieties obtained via PGE either will not be regulated or will be regulated in a different manner than traditional transgenic plants regulated by government agencies (Kuzma and Kokotovich, 2011; Waltz, 2012). Authorities worldwide are analyzing this technology and determining how to regulate plants that have been modified by site-specific nucleases. It is likely that not all uses of this technology will be regulated in the same way. For example, gene knockouts do not necessarily imply the introduction of exogenous DNA in a vegetable genome; rather, DNA sequences are generally deleted. Applications such as whole or partial target gene substitution can be used to create a wide spectrum of alterations. However, if the alteration of a single nucleotide is introduced in a plant gene and this is the only modification in the genome, should the plant be regulated? Traditional methods for the induction of mutations, such as radiation or chemicals, create similar results; however, plants produced by these methodologies are not regulated. Changes in regulatory policies will have important economic consequences because the costs of the deregulation of a GM plant are very high, and only large multinational companies normally have the financial and infrastructural resources to meet complex regulatory processes. Owing to high deregulation costs, only GM plants that bring considerable financial returns to a breeder are produced. Moreover, public enterprises, which normally have fewer resources, are limited in their capacity to develop GM plants that do not bring great financial return but meet the necessities of the producer.

Technological barriers that today limit or impair the use of PGE will likely disappear in the next few years, although this possibility does not guarantee that PGE will expand or replace current plant-breeding approaches. There are several special external factors for each species that might limit the applications of this technology. For example, PGE is more readily applicable in species with sequenced genomes to facilitate the identification of targets for DNA sequence modification. However, with the speed at which NGS technologies have been developed, for species without reference genomes, there will soon appear a considerable volume of genetic information to be employed in PGE.

Another limitation is that most PGE applications require efficient plant transformation methods. Therefore, cultivars with a greater research history and better-explored genetic resources (e.g., mutant stock) can better predict the impact of specific DNA modifications on the phenotype.

These resources are not available for many cultivated vegetable species. Additionally, efficient genetic transformation protocols have not yet been developed for all species for which these resources are available. Even when a cultivated species is transformable and has a sequenced genome, other limitations, such as a lack of information on functional genomics in the species, can hamper the prediction of phenotypic effects when considering PGE-induced modifications. In species with few functional genomic tools, only some genes will have been characterized, and it will be difficult to determine which target genes are responsible for a specific characteristic. Only a few candidate genes will be of interest, and only the most promising genes will compensate for the necessary investment in the target modification. The functional characterization of orthologous genes in associated model species can offer the best evaluation of candidates for many cultivated species.

Despite these difficulties, PGE remains an area of great interest in advanced genetic research. Not very long ago, modifying a specific nucleotide to create a desired phenotype was a visionary wish of breeders and geneticists. Given the incredible speed of the evolution of the techniques (e.g., GE, NGS, plant transformation, high-throughput phenotyping, etc.) used in the study of plant genetics, this vision is increasingly becoming reality. Until recently, scientists primarily used ZFN and TALEN technology to engineer plants precisely. However, the use of CRISPR technology in plants has been reported recently. In early 2013, scientists at the Academy of Sciences' Institute of Genetics and Developmental Biology in Beijing disabled four rice genes, suggesting that this technique could be used to engineer this crucial food crop (Shan et al., 2013). In wheat, they knocked out a gene that, when disabled, may enable resistance to powdery mildew. The low cost and ease of the CRISPR technology is enabling rapid advances to be made in its use.

In 2013, the CRISPR/Cas system was successfully applied to efficient genome editing in many plant species such as *Arabidopsis* and tobacco, dicot model plants (Nekrasov et al., 2013; Shan et al., 2013; Li et al., 2013; Feng et al., 2013; Jiang et al., 2013), and in rice, wheat and sorghum, monocot crop plants (Shan et al., 2013; Feng et al., 2013; Jiang et al., 2013; Miao *et al.*, 2013), suggesting that this technique can be used to engineer both dicot and monocot plants.

The initial CRISPR genome-editing reports in the literature all relied on DNA cutting, but other applications are rapidly appearing. For example, CRISPRi (Qi *et al.*, 2013), which, like RNAi, turns off genes in a reversible fashion, should be useful for studies of gene function. Using this technology, Cas9 was modified so that it and the associated guide RNA would still home in on a target but would not cut the DNA. In bacteria, the presence of Cas9 alone is enough to block transcription, but for mammalian applications, Qi and colleagues added a section of protein that represses gene activity. Its guide RNA is designed to target the regulation of gene promoter elements that immediately precede the gene target.

References

Arnould, S.; Delenda, C.; Grizot, S.; *et al.* 2011. The I-CreI meganuclease and its engineered derivatives: applications from cell modification to gene therapy. Protein Engineering, Design and Selection, 24: 27–31.

Bibikova, M.; Beumer, K.; Trautman, J.K.; Carroll, D. 2003. Enhancing gene targeting with designed zinc finger nucleases. Science, 300: 764.

Boch, J.; Bonas, U. 2010. Xanthomonas AvrBs3 family-type III effectors: discovery and function. Annual Review of Phytopathology, 48: 419–436.

Boch, J.; Scholze, H.; Schornack, S.; *et al.* 2009. Breaking the code of DNA binding specificity of TAL-type III effectors. Science, 326: 1509–1512.

Bogdanove, A.J.; Schornack, S.; Lahaye, T. 2010. TAL effectors: finding plant genes for disease and defense. Current Opinion in Plant Biology, 13: 394–401.

Cermak, T.; Doyle, E.L.; Christian, M.; *et al.* 2011. Efficient design and assembly of custom TALEN and other TAL effector-based constructs for DNA targeting. Nucleic Acids Research, 39: e82.

Christian, M.; Cermak, T.; Doyle, E.L.; *et al.* 2010. Targeting DNA double-strand breaks with TAL effector nucleases. Genetics, 186: 757–761.

Cui X.; Ji D.; Fisher D.A.; *et al.* 2011. Targeted integration in rat and mouse embryos with zinc-finger nucleases. Nature Biotechnology, 29: 64–67.

Curtin S.J.; Zhang F.; Sander J.D.; *et al.* 2011. Targeted mutagenesis of duplicated genes in soybean with zinc-finger nucleases. Plant Physiology, 156: 466–473.

Deltcheva, E.; Chylinski, K.; Sharma, C.M.; *et al.* 2011. CRISPR RNA maturation by trans-encoded small RNA and host factor RNase III. Nature, 471, 602–607.

Engler C.; Gruetzner R.; Kandzia R.; Marillonnet S. 2009. Golden gate shuffling: a one-pot DNA shuffling method based on type IIs restriction enzymes. PLoS One, 4: e5553.

Feng, Z.; Zhang, B.; Ding, W.; *et al.* 2013. Efficient genome editing in plants using a CRISPR/Cas system. Cell Research, 1–4.

Gao, H.; Smith, J.; Yang, M.; *et al.* 2010. Heritable targeted mutagenesis in maize using a designed endonuclease. Plant Journal, 61: 176–187.

Geurts, A.M.; Cost, G.J.; Freyvert, Y.; *et al.* 2009. Knockout rats via embryo microinjection of zinc-finger nucleases. Science, 325: 433.

Hafez, M.; Hausner, G. 2012. Homing endonucleases: DNA scissors on a mission. Genome, 55: 553–569.

Huang P.; Xiao A.; Zhou M.; *et al.* 2011. Heritable gene targeting in zebrafish using customized TALENs. Nature Biotechnology, 29: 699–700.

Jiang, W.; Zhou, H.; Bi, H.; *et al.* 2013. Demonstration of CRISPR/SgRNA /Cas9-mediated targeted gene modification in Arabidopsis, tobacco, sorghum and rice. Nucleic Acids Research, 1–12.

Jinek, M.; Chylinski, K.; Fonfara, I.; *et al.* 2012. A programmable dual-RNA-guided DNA endonuclease in adaptive bacterial immunity. Science (N.Y.), 337: 816–821.

Kay, S.; Bonas, U. 2009. How Xanthomonas type III effectors manipulate the host plant. Current Opinion in Microbiology, 12: 37–43.

Kim, Y.G.; Cha, J.; Chandrasegaran S. 1996. Hybrid restriction enzymes: zinc finger fusions to Fok I cleavage domain. Proceedings of the National Academy of Sciences U.S.A., 93: 1156–1160.

Kuzma, J.; Kokotovich, A. 2011. Renegotiating GM crop regulation. Targeted gene-modification technology raises new issues for the oversight of genetically modified crops. EMBO Reports, 12: 883–888.

Li, J.-F.; Norville, J.E.; Aach, J.; *et al.* 2013. Multiplex and homologous recombination-mediated genome editing in *Arabidopsis* and *Nicotiana benthamiana* using guide RNA and Cas9. Nature Biotechnology, 31, 688–691.

Li, L.; Piatek, M,J.; Atef, A.; *et al.* 2012. Rapid and highly efficient construction of TALE-based transcriptional regulators and nucleases for genome modification. Plant Molecular Biology, 78 (4–5): 407–416; doi: 10.1007/s11103-012-9875-4. Epub 2012 Jan 22.

Li, T.; Huang S.; Jiang W.Z.; *et al.* 2011. TAL nucleases (TALNs): hybrid proteins composed of TAL effectors and FokI DNA-cleavage domain. Nucleic Acids Research, 39: 359–372.

Li, T.; Liu B.; Spalding, M.H.; *et al.* 2012. High-efficiency TALEN-based gene editing produces disease-resistant rice. Nature Biotechnology, 30: 390–392.

Lloyd, A.; Plaisier, C.L.; Carroll, D.; Drews, G.N. 2005. Targeted mutagenesis using zinc-finger nucleases in Arabidopsis. Proceedings of the National Academy of Sciences U.S.A., 102: 2232–2237.

Maeder, M.L.; Linder, S.J.; Reyon, D.; *et al.* 2013. Robust, synergistic regulation of human gene expression using TALE activators. Nature Methods, 10: 243–245.

Maeder, M.L.; Thibodeau-Beganny, S.; Osiak A.; *et al.* 2008. Rapid "open-source" engineering of customized zinc-finger nucleases for highly efficient gene modification. Molecular Cell, 31: 294–301.

Mahfouz, M.M.; Li, L.; Shamimuzzaman, M.; *et al.* 2011. De novo-engineered transcription activator-like effector (TALE) hybrid nuclease with novel DNA binding specificity creates double-strand breaks. Proceedings of the National Academy of Sciences U.S.A., 108: 2623–2628.

Mahfouz, M.M.; Li, L.; Piatek, M.; *et al.* 2012. Targeted transcriptional repression using a chimeric TALE-SRDX repressor protein. Plant Molecular Biology, 78: 311–321.

Mali, P.; Yang, L.; Esvelt, K.M.; *et al.* 2013. RNA-guided human genome engineering via Cas9. Science, 339: 823–826.

Marton, I.; Zuker, A.; Shklarman, E.; *et al.* 2010. Nontransgenic genome modification in plant cells. Plant Physiology, 154: 1079–1087.

Mashimo, T.; Takizawa, A.; Voigt, B.; *et al.* 2010. Generation of knockout rats with X-linked severe combined immunodeficiency (X-SCID) using zinc-finger nucleases. PLoS One, 5: e8870.

Miao, J.; Guo, D.; Zhang, J.; *et al.* 2013. Targeted mutagenesis in rice using CRISPR-Cas system. Cell Research, 1–4.

Miller, J.C.; Tan, S.; Qiao, G.; *et al.* 2011. A TALE nuclease architecture for efficient genome editing. Nature Biotechnology, 29: 143–148.

Morbitzer, R.; Romer, P.; Boch, J.; Lahaye, T. 2010. Regulation of selected genome loci using de novo-engineered transcription activator-like effector (TALE)-type transcription factors. Proceedings of the National Academy of Sciences U.S.A., 107: 21617–21622.

Moscou, M.J.; Bogdanove, A.J. 2009. A simple cipher governs DNA recognition by TAL effectors. Science, 326: 1501.

Nekrasov, V.; Staskawicz, B.; Weigel, D.; *et al.* 2013. Targeted mutagenesis in the model plant *Nicotiana benthamiana* using Cas9 RNA-guided endonuclease. Nature Biotechnology, 31, 691–693.

Petolino, J.F.; Worden, A.; Curlee, K.; *et al.* 2010. Zinc finger nuclease-mediated transgene deletion. Plant Molecular Biology, 73: 617–628.

Porteus, M.H.; Baltimore, D. 2003. Chimeric nucleases stimulate gene targeting in human cells. Science, 300: 763.

Qi, L.S.; Larson, M.H.; Gilbert, L.A.; *et al.* 2013. Repurposing CRISPR as an RNA-guided platform for sequence-specific control of gene expression. Cell, 152 (5): 1173–1183; doi: 10.1016/j.cell.2013.02.022.

Ramirez, C.L.; Foley, J.E.; Wright, D.A.; *et al.* 2008. Unexpected failure rates for modular assembly of engineered zinc fingers. Nature Methods, 5: 374–375.

Reyon, D.; Tsai, S.Q.; Khayter, C.; *et al.* 2012. FLASH assembly of TALENs for high-throughput genome editing. Nature Biotechnology, 30: 460–465.

Sander, J.D.; Cade, L.; Khayter, C.; *et al.* 2011a. Targeted gene disruption in somatic zebrafish cells using engineered TALENs. Nature Biotechnology, 29: 697–698.

Sander, J.D.; Dahlborg, E.J.; Goodwin, M.J.; *et al.* 2011b. Selection-free zinc-finger-nuclease engineering by context-dependent assembly (CoDA). Nature Methods, 8: 67–69.

Sanjana, N.E.; Cong, L.; Zhou, Y.; *et al.* 2012. A transcription activator-like effector toolbox for genome engineering. Nature Protocols, 7: 171–192.

Shan, Q.; Wang, Y.; Li, J.; *et al.* 2013. Targeted genome modification of crop plants using a CRISPR-Cas system. Nature Biotechnology, 31: 686–688.

Shukla, V.K.; Doyon, Y.; Miller, J.C.; *et al.* 2009. Precise genome modification in the crop species Zea mays using zinc-finger nucleases. Nature, 459: 437–441.

Stoddard, B.L. 2011. Homing endonucleases: from microbial genetic invaders to reagents for targeted DNA modification. Structure, 19: 7–15.

Taylor, G.K.; Petrucci, L.H.; Lambert, A.R.; *et al.* 2012. LAHEDES: the LAGLIDADG homing endonuclease database and engineering server. Nucleic Acids Research, 40: W110–W116.

Tesson L.; Usal C.; Menoret S.; *et al.* 2011. Knockout rats generated by embryo microinjection of TALENs. Nature Biotechnology, 29: 695–696.

Tranel, P.; Wright, T. 2002. Resistance of weeds to ALS-inhibiting herbicides: what have we learned? Weed Science, 50: 700–712.

Waltz, E. 2012. Tiptoeing around transgenics. Nature Biotechnology, 30: 215–217.

Wang, M.; Helliwell, C.; Wu, L.; *et al*. 2008. Hairpin RNAs derived from RNA polymerase II and polymerase III promoter-directed transgenes are processed differently in plants. RNA, 14: 903–913.

Wang, Y.; Yau Y.-Y.; Perkins-Balding, D.; Thomson, J.G. 2011. Recombinase technology: applications and possibilities. Plant Cell Reports, 30 (3): 267–285.

Wiedenheft, B.; Sternberg, S.H., Doudna, J.A. 2012. RNA-guided genetic silencing systems in bacteria and archaea. Nature, 482: 331–338.

Wright, D.A.; Thibodeau-Beganny, S.; Sander, J.D.; *et al*. 2006. Standardized reagents and protocols for engineering zinc finger nucleases by modular assembly. Nature Protocols, 1: 1637–1652.

Wright, D.A.; Townsend, J.A.; Winfrey, R.J., Jr.,; *et al*. 2005. High-frequency homologous recombination in plants mediated by zinc-finger nucleases. Plant Journal, 44: 693–705.

Zhang, F.; Cong, L.; Lodato, S.; *et al*. 2011. Efficient construction of sequence-specific TAL effectors for modulating mammalian transcription. Nature Biotechnology, 29: 149–153.

Zhang F.; Maeder M.L.; Unger-Wallace E.; *et al*. 2010. High frequency targeted mutagenesis in *Arabidopsis thaliana* using zinc finger nucleases. Proceedings of the National Academy of Sciences U.S.A., 107: 12028–12033.

Zhang Y.; Zhang F.; Li X.; *et al*. 2013. Transcription activator-like effector nucleases enable efficient plant genome engineering. Plant Physiology, 161: 20–27.

11 RNA Interference

Francisco J.L. Aragão,[a] Abdulrazak B. Ibrahim,[a,b,c] and Maria Laine P. Tinoco[a]

[a]Embrapa Genetic Resources and Biotechnology, Brasília, DF, Brazil
[b]Department of Biochemistry, Ahmadu Bello University, Zaria, Kaduna, Nigeria
[c]Department of Cell Biology, University of Brasilia, DF, Brazil

Introduction

The second half of the 20th century brought with it great advances in biology, which allowed scientists to better understand biochemical pathways in living organisms and develop methods of manipulating these pathways with a high degree of precision in attempts to either unveil the mysteries behind such pathways or address biological problems for optimal utilization of natural resources in agriculture, health, and industry. In agriculture, as in other sectors, the development of recombinant DNA technology in the 1970s ushered in a new era and the emergence of genomic and proteomic tools, which could compliment already existing traditional methods of breeding, to counter both old leading to new biotic and abiotic stresses that hamper agricultural productivity. Among the several strategies that rely on genetic engineering, post-transcriptional gene silencing (PTGS), or RNA interference (RNAi), stands out as a method of choice for its practicality and specificity. The search for plants with resistance against viruses led to the accidental discovery of the phenomenon of PTGS, paving the way for the development of RNAi models (Angell and Baulcombe, 1997). The manipulation of naturally occurring gene silencing pathways in the laboratory has led to the generation of genetically modified plants capable of suppressing the expression of endogenous genes and invasive nucleic acids (for a review on this see Souza *et al.*, 2007; Aragão and Figueiredo, 2008).

In recent decades, few concepts in biotechnology have been th subjects of greater advances in terms of practical applications than RNAi. Indeed the experimental demonstration that has led to the comprehension of the mechanisms involved in gene silencing mediated by RNA represents an important milestone in understanding the biological function of genes. This has opened new avenues for understanding biological systems and

Omics in Plant Breeding, First Edition. Edited by Aluízio Borém and Roberto Fritsche-Neto.
© 2014 John Wiley & Sons, Inc. Published 2014 by John Wiley & Sons, Inc.

serves as a powerful tool for studying interaction amongst organisms, development of elite varieties for agriculture, and design and development of therapeutic agents for human health. Additionally, RNAi techniques are relevant in studies involving the search for improved nutritional values in plants, and in the development of plants that are better adapted to different ecosystems, as well as optimal utilization of raw materials derived from plants for industrial use.

Discovery of RNAi

The existence of naturally occurring gene silencing phenomena in organisms as diverse as viruses, fungi, plants, and animals is a clear indicator that it is an evolutionary stable strategy. Although gene silencing strategies have been shown to be much more efficient in viruses, RNAi technologies that rely on the phenomenon have been widely applied in pests and pathogens such as bacteria, fungi, nematodes, and insects.

One of the pioneer experiments involving the application gene silencing strategy was reported in 1986 when workers demonstrated that plants could be genetically engineered to exhibit resistance against viral diseases (Abel *et al.*, 1986). In this experiment, a chimeric gene containing the coat protein gene (CP) of *Tobacco mosaic virus* (TMV) was introduced into cells of *Nicotiana tabacum* via *Agrobacterium tumefaciens*. The plants regenerated from the transformed cells expressed the CP gene, and when inoculated with TMV showed delayed development of symptoms. Indeed 10–60% of the plants showed no symptoms at all. In another experiment, transgenic plants transformed to express a complementary RNA sequence (antisense RNA) of the coat protein gene of TMV were protected when challenged with the virus. It was further demonstrated that the accumulation of antisense RNA was responsible for this protection (Powell *et al.*, 1989). Although these pioneer experiments showed that the presence of viral RNAs resulting from transgene expression was responsible for the viral resistance observed in the plants, the mechanisms of resistance involved were not fully understood at the time.

Experiments conducted by Napoli *et al.* (1990) helped in elucidating the mechanism of endogenous gene silencing. By introducing the gene for chalcone synthase in petunia, Napoli *et al.* (1990) expected that the gene could be overexpressed thereby increasing pigmentation in flowers. In this way, plants may be generated with dark phenotype due to accumulation of anthocyanin. To their dismay, the introduced gene actually blocked the synthesis of anthocyanin and led to the generation of plants with white flowers. This phenomenon was referred to as co-suppression. Two years later, a similar phenomenon was observed in *Neurospora crassa* by Romano and Macino (1992). The fungus was transformed to super express the *albino-1* gene (*al*-1), involved in carotenoid biosynthesis, which confers an orange color to the fungus. However, the introduction of an extra copy of the *al*-1 gene produced colonies with the albino phenotype. This phenomenon was referred as quelling.

It was not until 1998 that the phenomenon was fully understood, and the term RNA interference was thus coined by Fire *et al.* (1998) in their now famous experiments with *Caenorhabditis elegans*. By injecting double-stranded RNAs (dsRNA) into nematodes, Fire *et al.* (1998) demonstrated that specific genes could be silenced at the post-transcriptional level. They further demonstrated that this silencing can indeed be "propagated" over a wide section of the nematode following injection of dsRNA into its extracellular abdominal cavity. The same effect was also observed when *C. elegans* were fed with *Escherichia coli* that transcribed recombinant dsRNA or indeed when the nematode was immersed in preparations containing dsRNA.

Mechanism of RNA Interference

RNAi evolved as a natural cellular defense mechanism against viruses, genomic confinement of retrotransposons, and as a cellular strategy for post-transcriptional regulation of gene expression. Knowledge of this mechanism has transformed into a technology used to silence specific genes leading to the creation of knock-out phenotypes in both transgenic plants through the production of sequence specific RNA hairpin, and by infection with recombinant RNA viruses harboring sequences of the target gene. Today, we know that co-suppression and virus-induced gene silencing share mechanistic similarities, thanks to biochemical studies conducted over the years. The pathway leading to gene silencing mediated by RNAi involves several steps, key among which is the generation of small RNA molecules *in vivo* (Figure 11.1).

A central feature in the mechanism of RNA inference is the participation of small RNA molecules. There are two types of small RNA molecules: small interfering RNAs (siRNA) and microRNA (miRNA). The process is initiated by an endonuclease RNase III known as dicer, which processes double stranded RNA generating small RNA molecules that range in size from 20–30 nucleotides which ultimately mediate the degradation of their complimentary RNAs (Angaji *et al.*, 2010; Czech and Hannon, 2011). The siRNA are processed by dicer-like enzymes (DCL2, DCL3, and DCL4) from a long double strand of RNA. On the other hand, DCL1 processes the precursors of miRNA exported from the nucleus (Xei *et al.*, 2004) (Figure 11.1).

Following dsRNA processing, siRNAs are assembled unto a multicomponent nuclease known as RNA induced silencing complex (RISC) (Hammond *et al.*, 2000; Figure 11.1). Originally identified by

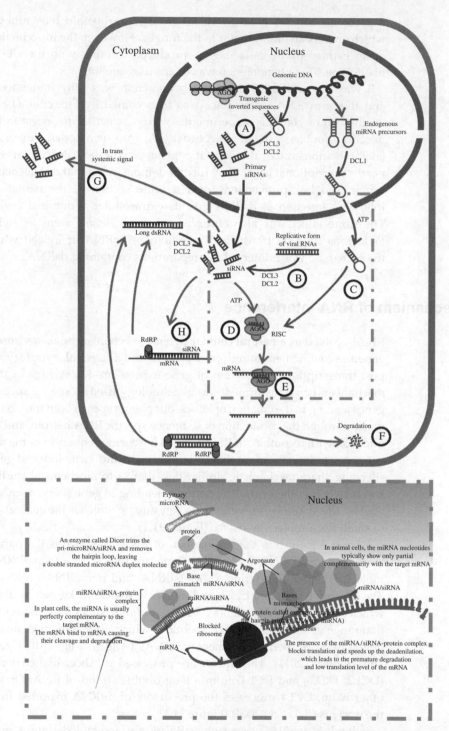

Figure 11.1   Gene silencing pathway. Dicer-like proteins processing transcripts containing inverted sequences (A), derived from viral RNA replication (B), and precursors of miRNA exported from the nucleus (C). Formation of siRNAs/RISC complex (D) directed to target RNA (E), which is subsequently, degraded (F); systemic silencing (G); and amplification by RdRP (H). (Source: Based on Souza *et al.*, 2007; Aragão and Figueiredo, 2008). (See color figure in color plate section).

fractionating an extract of specific nuclease from *Drosophila melanogaster* (Hammond *et al.*, 2001), RISC is a member of the Argonaut family. It is responsible for directing and cleaving of specific sequence of RNA in the cell (Martinez and Tuschul, 2004; Czech and Hannon 2011). This is achieved by cleaving the target mRNA at complimentary region of ten nucleotides upstream of a 5' residue of the RNA. A helicase in the RISC complex unwinds the siRNA duplex, pairing it with the antisense strand of the target mRNA, which, on its part, has a high degree of complementarity with the siRNA sequence. The cleavage leads to gene silencing by preventing the protein synthesis machinery from reading the mRNA, resulting in its degradation (Tolia and Joshua-Tor, 2006).

siRNAs are classified into primary and secondary siRNAs. While the primary siRNAs are generated through the activity of dicer, secondary siRNAs arise from an alternative pathway, which involves the activity of RNA dependent RNA polymerase (Pak and Fire, 2007). It appears that secondary siRNAs regulate gene expression involving signal transduction where they initiate the process of RNAi in the absence of the original signal for RNAi (Figure 11.1).

MircoRNAs are endogenous RNA molecules and play an important regulatory role during mRNA cleavage and repression of translation. They constitute one of the most abundant classes of regulatory molecules in multicellular organisms (Aukerman and Sakai, 2003; Bartel, 2004). In plants, miRNAs have been implicated in the control of cell division, leaf and meristematic patterning, environmental responses, heterochromatin maintenance, embryogenesis and development of meristem, leaves, anthers ,and vascular system (Palatnik *et al.*, 2003; Vazques *et al.*, 2004; Jover-Gil *et al.*, 2005).

For the formation of the primary miRNA, a transcript of a primary micro-RNA (pre-miRNA) synthesized from the introns of the RNA polymerase II enzyme gene is processed in the nucleus by a protein complex containing a ribonuclease specific to the double-strand producing an intermediary hairpin with 70 nucleotides. This pre-miRNA is then transported to the cytoplasm where it is cleaved by dicer. Following separation of the duplex strands, single stranded miRNA is incorporated into RISC forming the complex that inhibits translation or induces the degradation of target mRNA (Angaji *et al.*, 2010) (Figure 11.1).

Applications in Plant Breeding: Naturally Occurring Gene Silencing and Modification by Genetic Engineering

Over the years, interest in the use of RNAi mechanisms in plant breeding has been on the increase particularly due to the specificity and efficiency of the technique. Several crops have been used as targets of this technique to improve different characteristics of plants of agronomic importance. Efforts from various laboratories in research centers around the world have been rewarded with remarkable success. Such attempts have, in some cases, led to the development and commercialization of plants with improved agronomic traits. With advances in the post genomic era and the availability of high-throughput techniques, which have allowed for the generation of omics data for several species, the number of successful examples of genetically modified plants derived using RNAi technology has increased significantly (Sunilkumar *et al.*, 2006; Kusaba *et al.*, 2003; Bonfim *et al.*, 2007, Wang, *et al.*, 2011) (Table 11.1).

The application of RNAi techniques for improving plants evolved from studies of mutation in different plant species, which often led to the accidental discovery of RNA hairpin structures. The most common examples of such phenomenon have been observed in plants that display easily discernible phenotypic changes, such as seed and flower color. One such example is the change in color of the seed coat of soybean (*Glycin max*). In the plant, seed coat color is determined by the accumulation of anthocyanins. A key enzyme in the biosynthetic pathway of anthocyanins (besides other secondary metabolites such as isoflavones) is chalcone synthase (CHS) (Palmer *et al.*, 2004). At the chromosomal level, control of pigmentation is mediated by four alles (I, i^i, i^k, i) of locus I (inhibitor). Of these, I, i^i, and i^k are dominant alleles where I is responsible for the phenotypic features when seeds are colorless or bear yellow coloration, i^i gives rise to pigmented husk, and i^k gives rise to seeds with patches of pigment. In contrast, the i allele is recessive and produces seeds with brown or black pigment (Todd and Vodkin, 1993). Structural studies of the I locus (located on chromosome 8) revealed two inverted repeat clusters on the genes *CHS1*, *CHS3*, and *CHS4* (Todd and Vodkin, 1996; Tuteja and Vodkin, 2008). Six other CHS coding genes (*CHS2*, *CHS5*, *CHS6*, *CHS7*, *CHS8*, and *CHS9*) are also found in soybean and varieties with colorless seeds have reduced transcript level of *CHS* (Tuteja *et al.*, 2004). Subsequent studies further reported having found large quantities of siRNAs (predominantly 22nt), which corresponded to the regions of the *CHS* genes (Figure 11.2). These small RNA molecules are indeed specific to seed coat and arise from the transcription of *CHS1*, *CHS3*, and *CHS4* arranged in inverted repeat regions, leading to the formation of dsRNA (Tuteja *et al.*, 2009). Similarly, when *C2-Idf* allele (*colorless2*; containing a mutated chalcone synthase gene) occurs in the homozygous state, different seed parts are colorless (pericarp, aleurone layer of the endosperm, and vegetative organs). Plants with functional heterozygous *C2* allele exhibit an intermediary phenotype, characterized by lesser pigmentation (Vedova *et al.*, 2005). Cloning and sequence analysis of *C2-Idf* allele showed that its structure is quite different from the normal *C2* allele as two of its

Table 11.1 Examples of crops engineered using RNAi technology.

Crop	Target gene	Strategy	Application	Reference
Apple	Mal d 1	Silencing of the gene coding for Mal d 1	Development of apple free of the allergen Mal d 1	Gilissen et al., 2005
Banana	rep gene of BBTV	Transformation using an RNAi vector with hpRNA of rep gene of Banana bunchy top virus (BBTV)	Development of banana resistant to Banana bunchy top virus	Wang, Abbott, and Waterhouse, 2000
Barley	Sequence of BYDV-PAV	Transformation of plants with RNAi vector with hpRNA of BYDV-PAV	Development of plants resistant to BYDV	Ogita et al., 2003
Coffee	CaMXMT1	Silencing of the gene coding for 7-N-methylxanthine methyltransferase	Development of plants with reduced caffeine content	Sunilkumar et al., 2006
Cotton	δ-cadinene synthase	Silencing of the gene for δ-cadinene synthase in seeds	Reduction in gossypol	Yin et al., 2007
Ground nut	FAD2	Use of RNAi to regulate oleate desaturase (FAD2)	Increase in oleic acid	Tang et al., 2004
Maize	DHPS	Silencing of zein	Reduction in the catabolism of lysine (accumulation of lysine) and improved seed germination	Hily et al., 2005
Plum	CP-PPV	Silencing of the gene for CP-PPV during germination	Development of plants resistant to Plum pox virus (PPV)	Allen et al., 2004
Poppy	COR	Silencing of the gene coding for codeinone reductase (COR)	Elimination of narcotic alkaloids (morphine)	Kusaba et al., 2003
Rice	Multigene family of lgc	Plants with low glutenin Content-1	Nutrition for patients with celiac disease	Nunes et al., 2006
Soybean	GmMIPS	Silencing of the gene coding for the enzyme myoinositol-1-phosphate synthase (GmMIPS)	Reduction in the level of phytates	Shimada et al., 2006
Sweet potato	SBEII	Silencing of the gene for the protein SBE, which generates starch with branches	Increase in the level of amylase	Davuluri et al., 2005
Tomato	DET1	Suppression of the expression of DET1	Increase in the levels of carotenoids and flavonoids in fruits without affecting growth regulators	Le et al., 2006
Tomato	Lyc e 1.01	Silencing of the gene coding for profilin	Reduction in allergic reactions associated with profilin found in many fruits	Wang, Abbott, and Waterhouse, 2000
Wheat	BYDV	Construction using Barley yellow dwarf virus (BYDV) sequences	Immune lines resistant to virus	

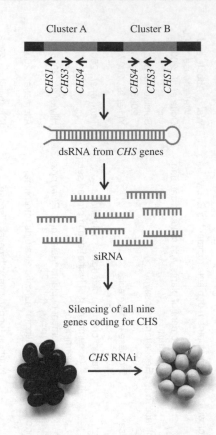

Figure 11.2 **Silencing of *CHS* genes coding for chalcone synthase (a key enzyme in the biosynthesis of anthocyanins) in soybean. The presence of inverted repeat sequences of *CHS1*, *CHS3*, and *CHS4* leads to formation of RNA hairpin (dsRNA), which is processed to form siRNA, leading to silencing of all of the nine CHS genes. This manifests in the phenotypic characteristics of the seeds as colorless (yellow). (See color figure in color plate section).**

three copies of the *CHS* gene lay side by side in an inverted orientation, leading to reduction in the level of its mRNA and consequently the enzyme (Dooner, 1983; Franken *et al.*, 1991). Indeed siRNAs have been found in plants containing C2-*Idf* allele and not in normal homozygous containing C2, indicating that the colorless phenotype is mediated by RNAi (Vedova *et al.*, 2005).

Another well characterized example is seen in rice with reduced levels of glutenin. The consumption of food substances with reduced levels of glutenin is important in patients with celiac disease whose diet must not contain this protein. The phenotype with a low level of glutenin is generated through an RNAi mechanism as a result of two inverted copies of genes near the glutenin coding gene on *Lgc1* locus (Kusaba *et al.*, 2003).

A number of techniques that rely on RNAi pathways have been widely developed and applied in order to knock out genes of interest in different plants with a view to unlocking the full agronomic potentials of the crops under physical and biological conditions that would otherwise make such feats impossible. These strategies often seek to improve productivity, confer resistance and/or tolerance to many pests and diseases (Wang *et al.*, 2000; Bonfim *et al.*, 2007).

Furthermore, the technique has been used to improve the nutritional value of plants, in addition to optimizing use of raw materials derived from plants for industrial use (Ossowski *et al.*, 2008; Sunilkumar *et al.*, 2006; Yin *et al.*, 2007; Shimada *et al.*, 2006).

A key component of the strategies involved in developing genetically modified plants in this respect is the use of RNAi vectors. Several of such vectors are now available and a representation of chief among them is given in Figure 11.3. In rice for example, the expression vector pNW55, derived from a natural miRNA (osa-MIR528 of rice) in which an artificially inserted miRNA sequence, was designed to silence *Pds*, *Spl11*, and *Eui1/CYP714D1* (Warthmann *et al.*, 2008). A similar approach had earlier been employed to silence the *P69* gene of *Turnip yellow mosaic virus* and the HC-pro gene of *Turnip mosaic virus* in *Arabidopsis thaliana* by using miR159 from the plant to construct a vector that expressed artificial miRNAs (amiRNA). Plants generated from this work were reported to be resistant to the two viruses even under low temperature (15 °C) (Niu *et al.*, 2006). This approach is immensely important in attempts to silence endogenous genes of plants especially where complete genomes of such plants are available. However, in cases where the target organisms are pathogens (virus, fungus, nematodes), specificity constitutes a problem in attempts to confer wide and reliable resistance in both *cis* and *trans* approaches.

Resistance to Viruses

Several plants are resistant to viruses by virtue of an inherent dsRNA and siRNA generating system whose targets are gene sequences essential for viral pathogenicity. For example, siRNAs sharing 100% similarity with distinct genetic and intragenic regions of *Mungbean yellow mosaic India virus* (MYMIV), a begomovirus which causes yellow mosaic disease, have been observed in mungbean. In the resistant line PK416, siRNAs were found to correspond to an intragenic region (IR) of MYMIV, while in the susceptible lines, most of the siRNAs correspond to the genetic regions and are present in low concentrations. It was also demonstrated that the viral genomes in resistant plants were methylated in the intragenic regions (Yadav and Chattopadhyay, 2011).

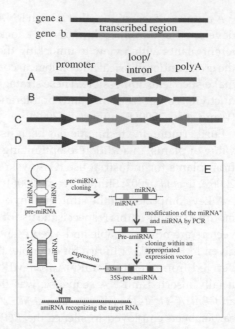

Figure 11.3 Vectors used in the stable transformation of plants are generally designed to produce hairpin structures (after transcription of RNA or dsRNA). Here, a transcribed sequence of a gene is amplified and placed under the control of a promoter in forward (sense) and reverse (antisense) directions spaced by an intron or a spacer region (loop) (A, B). Sequences of two or more genes can be used in the same expression cassette (C). It is also possible to join two expression cassettes harbouring gene fragments cloned in forward and reverse directions separated by a spacer (D). In (E), a vector designed to generate a modified miRNA by introducing a target gene sequence into a natural miRNA region (such as miR159 of *A. thaliana* and miR528 of *Oriza sativa*) is presented. The resulting vector expresses artificial miRNA (amiRNA) whose target may be the endogenous gene or that of an intracellular pathogen. (See color figure in color plate section).

Although RNAi mediated resistance to viruses is a natural phenomenon in plants, it is not effectively present in many productive lines because siRNA molecules identical to viral sequences usually appear at later stages of infection in some of these lines, when it is not stoichiometrically favorable to control the infection (Rodriguez-Negrete *et al.*, 2009; Aregger *et al.*, 2012). However, this can be circumvented by mimicking the mechanism using recombinant DNA technology to generate siRNAs that can confer resistance or immunity to plants against viruses even before the onset of infection. Currently, there are several reports of protocols in which RNAi strategies have been used to generate plants resistant to viruses based on either RNA or DNA genomes (Prins *et al.*, 2008; Runo, 2011; Prins, 2003; Vanderschuren *et al.*, 2007; Bonfim *et al.*, 2007; Aragão and Faria, 2009; Lucioli *et al.*, 2003; Fuentes *et al.*, 2006; Vanderschuren *et al.*, 2009; Hashmi *et al.*, 2011; Vanderschuren *et al.*, 2012).

However, the first report on gene silencing appeared in 1986 when tobacco plants were transformed with the coat protein gene of *Tobacco mosaic virus* (TMV) (Beachy *et al.*, 1987). Following this, more than 100 publications have appeared reporting on the development of genetically modified plants resistant to viruses of different groups.

At the commercial level, the first crop varieties resulting from this technology were tobacco resistant to *Tobacco mosaic virus* (TMV) in China, and papaya resistant to *Papaya ringspot virus* (PRSV), which has been in cultivation in the United States since 1998. All of these crops are now available to farmers. Other commercially available crops in the United States include pumpkins resistant to WMV, ZYMV, and CMV, and virus resistant potatoes.

However, the first deliberate transformation to express a dsRNA construct harbouring intron–hairpin RNA (hpRNA) was reported in 2000 using wheat in which gene sequences of a polymerase from *Barley yellow dwarf virus* (BYDV) were expressed. Plant lines arising from this were immune to the virus following tests using ELISA (Wang *et al.*, 2000). Shortly thereafter, transgenic tobacco plants expressing sense and antisense RNAs of DNA-A of *Cotton leaf curl virus* (CLCuV DNA-A) and DNA-B of CLCuV were generated. The siRNAs of DNA-A inhibited viral replication while those of DNA-B conferred resistance against CLCuV to the plants (Asad *et al.*, 2003). In addition, researchers generated tomato plants with resistance against *Tomato yellow leaf curl Sardinia virus* (TYLCSV) using RNA hairpin constructs containing truncated Rep protein gene of TYLCSV (Yang *et al.*, 2004).

In an attempt to extend this technique to leguminous plants, Poogin *et al.* (2003) used a dsRNA construct to silence the promoter sequence of DNA-A of *Vigna mungo yellow mosaic virus* (VMYMV), leading to the expression of dsRNA of a conserved region of VMYMV in *Vigna* spp., resulting in resistance against viral infection. Similarly, Bonfim *et al.* (2007) applied RNAi technology using a viral *AC1* gene sequence encoding a multifunctional protein (Rep) of the *Bean golden mosaic virus* (BGMV) to generate transgenic common bean (*Phaseolus vulgaris* L.) resistant to geminivirus. The choice of this viral gene (*AC1* or *Rep*) for the construction of the transformation vector was based on the fact that Rep protein plays an essential role in the viral infection cycle and as it is the only protein required for replication. The vector used was constructed from a DNA fragment of 411 bp of *AC1* gene of BGMV. This resulted in the development of an event now known as Embrapa 5.1: the first transgenic line approved for commercial use through the application of Brazilian technology and following Brazilian biosafety regulations set by the Brazilian Technical Biosafety Commission (CTNBio) (Aragão and Faria, 2009). This strategy can also be applied to combat other devastating diseases such as geminivirus attacking maize and cassava in Africa, and tomato worldwide.

Host-induced Gene Silencing

The discovery that genetically modified plants can be used to control pathogenic organisms when engineered to release siRNA specific to a vital gene in susceptible pathogens is another indication of the great potentials that RNAi techniques can unlock. Such a feat was reported by Tinoco *et al*. (2010) who demonstrated *in vivo* interference using the pathogenic fungus *Fusarium verticillioides*. In their experiments, inoculation of mycelium in transgenic tobacco plants, engineered to express siRNA from a dsRNA corresponding to a transgene, specifically silenced genes in the fungus. This proved a powerful tool for understanding the molecular interaction between plants and pathogens and symbiotic interactions. From the viewpoint of biotechnology, silencing fungal genes by siRNAs generated by host plant represents an important strategy for developing fungal resistance in plants and other organisms Koch *et al*., 2013). This movement of silencing signals in the form of siRNAs derived from one organism, exerting their effects on another, has also been observed in nematodes (Waterhouse, Graham, and Wang, 1998). These workers reported that gene silencing was triggered when nematodes were fed on a diet made from transgenic plants engineered to express dsRNA. The same phenomenon was observed in herbivorous insects fed with transgenic plants expressing dsRNAs of genes that are vital to insects (Baum *et al*., 2007; Mao *et al*., 2007).

In 2007 Roney, Khatibi, and Weswood reported on the systemic movement of mRNA through phloem between tomato and the parasitic plant *Cuscuta pentagona* Engelm. Experiments described by Tomilov *et al*. (2008) also showed that host plants transformed with constructs that generate interference hairpins can silence the expression of target gene in parasitic plants. Roots of transgenic *Triphysaria versicolor* expressing the reporter gene *gus* became parasitic to transgenic lettuce expressing RNA hairpin containing a fragment of the *gus* gene (hpGUS). Additionally, Aly *et al*. (2009) showed that a construct containing the binary vector pBIN-IR-M6PR inserted in the tomato genome can silence the expression of the *M6PR* gene in tubers of *Orobanche* that parasitize the roots of transgenic plants. The observation that molecules produced by host plants are responsible for silencing specific genes in parasitic plants suggests a new strategy for engineering plants resistant to parasites.

Insect and Disease Control

Although commercial biotechnology has made available protocols for the control of diseases transmitted by both Coleopteran and Lepidopteran insects through the expression of insecticidal protein from *Bacillus thuringiensis* (Bt toxin), the emergence of resistance to Bt toxin in some insect biotypes underscores the need to develop new control strategies that require a different mode of action (Baum *et al.*, 2007). Silencing of essential insect genes mediated by dsRNA can interrupt feeding or lead to death of susceptible insects. In this respect, it has been demonstrated that ingestion of RNAs provided in an artificial diet induces RNA interference in Coleopterans such as *Diabrotica* sp. (Gordon and Waterhouse, 2007; Baum *et al.*, 2007; Gatehouse, 2008; Upadhyay *et al.*, 2011). The development of transgenic corn engineered to express dsRNAs against the V-ATPase of corn rootworm, which showed suppression of mRNA in the insect and reduction in feeding damage, is a powerful indicator that the RNAi pathway can be exploited to control insect pests in plants by expression of a dsRNA *in vitro* (Baum *et al.*, 2007). Similarly, transgenic cotton and *Arabidopsis* plants engineered to express dsRNA directed against Cyt P450, a detoxification enzyme (coded for by CYP6AE14) for gossypol in cotton bollworm, induced feeding damage in insects (Mao *et al.*, 2007).

Indeed the fact that the RNAi machinery is present in all living insects further highlights the potentials for the use of this approach for insect control by interrupting the expression of their essential genes. This is possible even for insect species that lack a systemic RNAi response because genes expressed in insect midgut are susceptible to silencing by dsRNA when ingested in a diet (Huvenne and Smagghe, 2010).

Improving Nutritional Values

Although many plants may be regarded as sources of proteins, a good number of them are deficient in certain essential amino acids or, when present, these important nutrients are accumulated in cellular compartments that make their utilization difficult or indeed toxic for human and animal consumption. Accordingly, various breeding programs seek to increase levels of amino acids in order to add value to crops and make such nutrients bioavailable (Tu, Godfrey, and Sun, 1998; Marcellin *et al.*, 1996). Among these amino acids are lysine and sulfur containing amino acids. For example, while a high level of lysine in seeds is beneficial, an increase in the level of this amino acid in vegetative tissues is undesirable because it may lead to abnormal growth or hamper flower development. The pathway for the biosynthesis of lysine is under tight regulation by a feedback inhibition mechanism in which the amino acid inhibits the activity of dihydrodipicolinate synthase (DHPS), the first enzyme in the committed step of lysine biosynthesis. It has been demonstrated that mutations in tobacco *DHPS* gene may cause its encoded DHPS lysine to become insensitive, leading to overproduction of lysine in all plant organs (Frankard *et al.*, 1992; Negrutiu *et al.*, 1984). The RNAi technique has thus been used to improve the germination of seeds of *Arabidopsis* by silencing DHPS (Zhu and Galili, 2003; Zhu and Galili, 2004; Tang, Galili, and Zhuang, 2007). The same approach was used in maize to increase the level of lysine in seeds by manipulating the gene for zein, a protein normally associated with low nutritional quality. Using RNAi constructs derived from a fragment of the 22 kDa zein gene, researchers generated a dominant opaque maize phenotype with a low level of zein, which corresponded to an increase in the level of lysine improving the plant's nutritional value and promoting seed germination (Segal *et al.*, 2003).

The RNAi technique has also been used in soybean to silence the gene of myoinositol-1-phosphate synthase (*GmMIPS*), a key enzyme in the biosynthesis of phytic acid in seeds. Phytates are anti-nutritional factors that chelate divalent minerals such as zinc, calcium, iron, and others, present in food, reducing nutritional value. Phytates are also eliminated in the feces and thus may pose environmental contamination. In order to generate soybean plants with silenced *GmMIPS1*, a vector was constructed (pMIPSGm) in which *GmMIPS1* fragments were cloned in the reversed direction, generating sense and antisense arms. The resulting soybean plants showed partial silencing of this gene and led to the development of soybean lines with up to 94.5% reduction in phytates (Nunes *et al.*, 2006).

At the industrial level, potato (*Solanum tuberosum*) with high-amylose starch was developed using RNA interference to inhibit two genes coding for starch branching enzymes (*Sbe1* and *Sbe2*) resulting in transgenic lines with high-amylose, a quality desirable in the market (Shimada *et al.*, 2006).

Secondary Metabolites

Besides agronomic traits, industrial and pharmaceutical substances derived from plants can be enhanced using RNAi technology. This has been used to interfere with pathways of secondary metabolites in order to generate useful substances for pharmaceutical use and allelopathy. For example, gene silencing of codeinone reductase (COR) in opium (*Papaver somniferum*) led to accumulation of non-narcotic alkaloids (Allen *et al.*, 2004). In cotton, the technology was used to reduce the level of gossypol, a toxic compound that accumulates in seeds and restricts the use of cotton as a possible source of protein for humans. This was achieved by intervening in the gene expression of δ-cadinene synthase during seed development (Sunilkumar *et al.*, 2006).

Perspectives

Discovered less than 20 years ago, the RNAi mechanism has today become a powerful tool for understanding how genes function in various biological processes, thus constituting an important tool in metabolomics. Its application in the development of plants with farmer preferred agronomic traits has opened new opportunities hitherto unthinkable even with the best methods of classical breeding. Already, several technologies have been approved for commercialization in the United States and Brazil. With the explosion of knowledge on the biological and biochemical mechanisms underlying the RNAi pathway, our ability to fully harness and unlock the potentials of this mechanism, as both an experimental tool and a problem solving strategy, will undoubtedly increase. Despite such limitations as dependency on other techniques, which are sometimes not reproducible, RNAi technology will continue to be used alongside conventional breeding approaches for the development of new cultivars in the coming years. With advances in the development of tools for genetic manipulation of the plant genome, the coming years seem promising when it will be possible to effectively use strategies involving zinc-finger nucleases (ZFNs) (Isalan, 2012), TALENS (Sanjana *et al.*, 2012), and other endonucleases to generate new transformation events with remarkable success. This is even more so because with these strategies, it is possible to selectively mutagenize multiple gene copies resulting in precise silencing that could yield desirable phenotypes with a high degree of accuracy and requiring less time and resources for selection, molecular analyses, and biosafety tests.

Issues related to biosafety of the use of RNAi in genetically modified plants have been discussed extensively in some reviews (Hollingworth *et al.*, 2003; Petrick *et al.*, 2013; Parrott *et al.*, 2010). However, based on the evidence that humans, and indeed all animals, have been consuming foods with naturally occurring RNA molecules (including miRNA, siRNA, long dsRNA, and mRNA), it is reasonable to posit, in principle, that there is no reason to expect that consumption of genetically modified foods derived from RNAi could pose any health risks. Plants have an average of 1 mg of total RNA per mg of tissue (Ivashuta *et al.*, 2009; Lassek and Montag, 1990). Of this total, the non-coding RNA (tRNA, rRNA, antisense-ssRNA, dsRNA from external sources, such as viruses, miRNAs, and siRNAs) constitute the larger percentage. It is worth remembering that man has been feeding on animals (with a history of safe consumption) and these animals contain miRNA and siRNA with a high similarity to human genes (Carthew and Sontheimer, 2009; Petrick *et al.*, 2013; Jensen *et al.*, 2013). Despite this, however, it is extremely important that for each product generated using RNAi technology, rigorous biosafety analyses are conducted. Already, a number of genetically modified plants expressing siRNA have been commercially released following such regulations. These plants include: Flavr Savr™ tomato modified to silence the gene for polygalacturonase in fruits; pumpkin resistant to *Watermelon mosaic virus 2* and *Zucchini yellow mosaic virus*: and papaya resistant to *Papaya risgspot virus* (http://www.agbios.com).

Scientists will continue to rely on RNAi technology to discover and validate gene function, but more importantly, to generate desirable products in plants. The great advantage of the technique, symbolized by its specificity in sequence, tissue, and time of expression, allows for its relative ease of gene targeting with high silencing efficiency and potency, as against other methods that have higher tendencies for missing targets. Indeed, RNAi technology, by its very nature, has the ability to predict the effect of off-target silencing. As new generations of RNAi based transgenic crops emerge, further research is needed to meet the growing human need.

References

Abel, P.P.; Nelson, R.S. De B.; Hoffmann, N.; *et al.* 1986. Delay of disease development in transgenic plants that express the *Tobacco Mosaic Virus* coat protein gene. Science, 232: 738–743.

Allen, R.S.; Millgate, A.G.; Chitty, J.A.; *et al.* 2004. RNAi-mediated replacement of morphine with the non-narcotic alkaloid reticuline in opium poppy. Nature Biotechnology, 22: 1559–1566.

Aly, R.; Cholakh, H.; Joel, D.M.; *et al.* 2009. Gl-on: Gene silence of mannose 6-phosphate reductase in the parasitic weed *Orobanche aegyptiaca* through the production of homologous dsRNA sequences in the host plant. Plant Biotechnology Journal, 7: 487–498.

Angaji, S.A.; Hedayati, S.S.; Hosein, R.; *et al.* 2010. Application of RNA interference in plants. Plant Omics Journal, 3: 77–84.

Angell, S.M.; Baulcombe, D.C. 1997. Consistent gene silencing in transgenic plants expressing a replicating potato virus X RNA. The EMBO Journal, 16: 3675–3684.

Aragão, F.J.L; Figueiredo S.A. 2008. RNA interference as a tool for plant biochemical and physiological studies. In: Rivera-Domínguez, M., Rosalba-Troncoso, R., Tiznado-Hernández, M.E. (eds). A Transgenic Approach in Plant Biochemistry and Physiology, Kerala, India: Research Signpost, pp. 17–50.

Aragão, F.J.L.; Faria, J.C. 2009. Fist transgenic geminivirus-resistant plant in the field. Nature Biotecnology, 27: 1086–1088.

Aregger, M.; Borah, B.K.; Seguin, J.; *et al.* (2012) Primary and secondary siRNAs in geminivirus-induced gene silencing. PLoS Pathogens, 8 (9): e1002941.

Asad, S.; Haris, W.A.; Bashir, A.; *et al.* 2003. Transgenic tobacco expressing geminiviral RNAs are resistant to the serious viral pathogen causing cotton leaf curl disease. Archives of Virology, 148: 2341–2352.

Aukerman, M.J.; Sakai, H. 2003. Regulation of flowering time and floral organ identity by a microRNA and its APETALA2-like target genes. The Plant Cell, 15: 2730–2741.

Bartel, D.P. 2004. MicroRNAs: genomics, biogenesis, mechanism and function. Cell, 116: 281–297.

Baum, J.A.; Bogaert, T.; Clinton, W.; *et al.* 2007. Control of coleopteran insect pests through RNA interference. Nature Biotechnology, 25: 1322–1326.

Beachy, R.N.; Stark, D.M.; Deom, C. M.; *et al*. 1987. Expression of sequences of *Tobacco mosaic virus* in transgenic plants and their role in disease resistance. In: Tailoring Genes for Crop Improvement. Basic Life Sciences. New York: Plenum Publishing, vol. 41, pp. 169–180.

Bonfim, K.; Farias, J.C.; Nogueira, E.O.P.L.; *et al*. 2007. RNAi-mediated resistance to bean golden mosaic virus in genetically engineered common bean (Phaseolus *vulgaris*). Molecular Plant-Microbe Interactions, 20: 717–726.

Carthew, R.W.; Sontheimer, E.J. 2009. Origins and mechanisms of miRNAs and siRNAs. Cell, 136: 642–655.

Czech, B.; Hannon, G.J. 2011. Small RNA sorting: matchmaking for Argonautes. Nature Reviews Genetics, 12: 19–31.

Davuluri, G.R.; Tuinen, A.; Fraser, P.D.; *et al*. 2005. Fruit-specific RNAi mediated suppression of DET1 enhances carotenoid and flavonoid content in tomatoes. Nature Biotechnology, 23: 890–895.

Dooner, H.K.; Robbins, T.P.; Jorgensen, R.A. 1991. Genetic and developmental control of anthocyanin biosynthesis. Annual Review of Genetics, 25: 173–199.

Fire, A.; Xu, S.; Montgomery, M.K.; Kostas, S.A.; *et al*. 1998. Potent and specific genetic interference by double-stranded RNA in *Caenorhabditis elegans*. Nature, 391: 806–811.

Frankard, V.; Ghislain, M.; Jacobs, M. 1992. Two feedback-insensitive enzymes of the aspartate pathway in *Nicotiana sylvestris*. Plant Physiology, 99: 1285–1293.

Franken, P.; Niesbach-Klosgen, U.; Weydemann, U.; *et al*. 1991. The duplicated chalcone synthase genes C2 and Whp (white pollen) of *Zea mays* are independently regulated; evidence for translational control of Whp expression by the anthocyanin intensifying gene in. EMBO Journal, 10: 2605–2612.

Fuentes, A.; Ramos, P.; Fiallo, E.; *et al*. 2006. Intron–hairpin RNA derived from replication associated protein C1 gene confers immunity to *Tomato yellow leaf curl virus* infection in transgenic tomato plants. Transgenic Research, 15: 291–304.

Gatehouse, J.A. 2008. Biotechnological prospects for engineering insect-resistant plants. Plant Physiology, 146: 881–887.

Gilissen, L.J.W.J.; Bolhaar, S.T.H.; Matos, C.I.; *et al*. 2005. Silencing of the major apple allergen Mal 1 by using RNA interference approach. Journal of Allergy and Clinical Immunology, 115: 369–384.

Gordon, K.H.J.; Waterhouse, P M. 2007. RNAi for insect-proof plants. Nature Biotechnology, 25: 11.

Hammond, S.M.; Berntein, E.; Beach, D.; Hannon, G.J. 2000. An RNA-dicted nuclease mediates post-transcriptional gene silencing in Drosophila cells. Nature, 404: 293–296.

Hammond, T.M.; Caudy, A.A.; Hannon, G.J. 2001. Post-transcriptional gene silencing by double-strand RNA. Nature Reviews of Genetics, 2: 110–119.

Hashmi, J.A.; Zafar, Y.; Arshad, M.; *et al*. 2011. Engineering cotton (*Gossypium hirsutum* L.) for resistance to cotton leaf curl disease using viral truncated AC1 DNA sequences. Virus Genes, 42: 286–296.

Hily, J.-M.; Scorza, R.; Webb, K.; Ravelonandro, M. 2005. Accumulation of the long class of siRNA is associated with resistance to plum pox virus in a transgenic woody perennial plum tree. Molecular Plant-Microbe Interactions, 18: 794–799.

Hollingworth, R.M.; Bjeldanes, L.F.; Bolger, M.; *et al*. 2003. The safety of genetically modified foods produced through biotechnology. Toxicological Sciences, 71: 2–8.

Huvenne, H.; Smagghe, G. 2010. Mechanisms of dsRNA uptake in insects and potential of RNAi for pest control: A Review Journal of Insect Physiology, 56: 227–235.

Isalan, M. 2012. Zinc-finger nucleases: how to play two good hands. Nature Methods, 9: 32–34.

Ivashuta, S.I.; Petrick, J.S.; Heisel, S.E.; *et al*. 2009. Endogenous small RNAs in grain: semi-quantification and sequence homology to human and animal genes. Food and Chemical Toxicology, 47: 353–360.

Jensen P.D., Zang, Y., Wiggins, B.E., *et al*. 2013. Computational sequence analysis of predicted long dsRNA transcriptomes of major crops reveals sequence complementarity with human genes. GM Crops & Food, 4: 90–97

Jover-Gil, S.; Candela, H.; Ponce, M.R. 2005. Plant microRNAs and development. International Journal of Developmental Biology, 49: 733–744.

Koch, A.; Kumar, N.; Weber, N.; *et al*. 2013. Host-induced gene silencing of cytochrome P450 lanosterol C14α-demethylase–encoding genes confers strong resistance to *Fusarium* species Proceedings of the National Academy of Sciences U.S.A., published ahead of print.

Kusaba, M.; Miyahara, K.; Iida, S.; *et al*. 2003. Low glutelin content 1: a dominant mutation that suppresses the glutelin multigene family via RNA silencing in rice. The Plant Cell, 15: 1455–1467.

Lassek, E.; Montag, A. 1990. [Nucleic acid components in carbohydrate-rich food]. Zeitschrift fur Lebensmittel-Untersuchung und -Forschung, 190: 17–21.

Le, L.Q.; Lorenz, Y.; Scheurer, S.; *et al*. 2006. Design of tomato fruits with reduced allergenicity by dsRNAi-mediated inhibition of ns-LTP (Lyc e 3) expression. Plant Biotechnology Journal, 4: 231–242.

Lucioli, A.; Noris, E.; Brunetti, A.; *et al*. 2003. Tomato yellow leaf curl Sardinia virus Rep-derived resistance to homologous and heterologous geminiviruses occurs by different mechanisms and is overcome if virus-mediated transgene silencing is activated. Journal of Virology, 77: 6785–6798.

Marcellin, L.H.; Neshich, G.; Sá, M. F.; *et al*. 1996. Modified 2S albumins with improved tryptophan content are correctly expressed in transgenic tobacco plants. FEBS Letters, 385: 154–158.

Mao, Y.B.; Cai, W.J.; Wang, J.W.; *et al*. 2007. Silencing a cotton bollworm P450 monooxygenase gene by plant-mediated RNAi impairs larval tolerance of gossypol. Nature Biotechnology, 25: 1307–1313.

Martinez, J.; Tuschul, T. 2004. RISC is a 5′ phosphomonoester producing RNA endonuclease. Genes & Development, 18: 975–980.

Napoli, C.; Lemieux, C.; Jorgensen, R. 1990. Introduction of a chimeric chalcone synthase gene into petunia results in reversible co-suppression of homologous genes in trans. Plant Cell, 2: 279–289.

Negrutiu, I.; Cattoir-Reynearts, A.; Verbruggen, I.; Jacobs, M. 1984. Lysine overproducer mutants with an altered dihydrodipicolinate synthase from protoplast culture of Nicotiana sylvestris (Spegazzini and Comes) Theoretical and Applied Genetics, 68: 11–20.

Niu, Q.W.; Lin, S. S.; Reyes, J.L.; *et al.* 2006. Expression of artificial microRNAs in transgenic *Arabidopsis thaliana* confers virus resistance. Nature Biotechnology, 24: 1420–1428.

Nunes, A.C.S.; Vianna, G.R.; Cuneo, F.; *et al.* 2006. RNAi-mediated silencing of the myo-inositol-1-phosphate synthase gene (GmMIPS1) in transgenic soybean inhibited seed development and reduced phytate content. Planta, 224: 125–132.

Ogita, S.; Uefuji, H.; Yamaguchi, Y.; *et al.* 2003. Production of decaffeinated coffee plants by genetic engineering. Nature, 423: 823.

Ossowski, S.; Schwab, R.; Weigel, D. 2008. Gene silencing in plants using artificial microRNAs and other small RNAs. The Plant Journal, 53: 674–690.

Pak, J.; Fire, A. 2007. Distinct populations of primary and secondary effectors during RNAi in *C. elegans*. Science, 315: 241–244.

Palatnik, J.F.; Allen, E.; Wu, X.; *et al.* 2003 Control of leaf morphogenesis by microRNAs. Nature, 425: 257–263

Palmer, R.G.; Pfeiffer, T.W.; Buss, G.R.; Kilen, T.C. 2004. Qualitative genetics. In: Boerma, H.G., Specht, J.E. (eds). Soybeans: Improvement, Production and Uses. Madison, WI: American Society of Agronomy, pp. 137–233.

Parrott, W.; Chassy, B.; Ligon, J.; *et al.* 2010. Application of food and feed safety assessment principles to evaluate transgenic approaches to gene modulation in crops. Food and Chemical Toxicology, 48: 1773–1790.

Petrick, J.S.; Brower-Toland, B.; Jackson, A.L.; Kier, L.D. 2013. Safety assessment of food and feed from biotechnology-derived crops employing RNA-mediated gene regulation to achieve desired traits: a scientific review. Regulatory Toxicology and Pharmacology, 66: 167–176.

Prins, M. 2003. Broad virus resistance in transgenic plants. Trends in Biotechnology, 21: 373–375.

Prins, M.; Laimer, M.; Noris, E.; *et al.* 2008. Strategies for antiviral resistance in transgenic plants. Molecular Plant Pathology, 9: 73–83.

Pooggin, M.M.; Shivaprasad, P.V.; Veluthambi, K.; Hohn, T. 2003. RNAi targeting of DNA virus in plants. Nature Biotechnology, 21: 131–132.

Powell, P.A.; Stark, D.M.; Sanders, P.R,. Beachy, R.N. 1989. Protection against tobacco mosaic virus in transgenic plants that express tobacco mosaic virus antisense RNA. Proceedings of the National Academy of Sciences U.S.A., 86: 6949–6952.

Rodríguez-Negrete, E.A.; Carrillo-Tripp, J.; Rivera-Bustamante, R.F. 2009. RNA silencing against geminivirus: complementary action of posttranscriptional gene silencing and transcriptional gene silencing in host recovery. Journal of Virology, 83: 1332–1340.

Romano, N.; Macino, G. 1992. Quelling: transient inactivation of gene expression in *Neurospora crassa* by transformation with homologous sequences. Molecular Microbiology, 6: 3343–3353.

Roney, J.K.; Khatibi, P.A.; Westwood, J.H. 2007. Cross-species translocation of mRNA from host plants into the parasitic plant dodder. Plant Physiology, 143: 1037–1043.

Runo, S.; Alakonya, A.; Machuka, J.; Sinha, N. 2011. RNA interference as a resistance mechanism against crop parasites in Africa: a 'Trojan horse' approach. Pest Management Science, 67: 129–136.

Sanjana, N.E.; Cong, L.; Zhou, Y.; *et al.* 2012. A transcription activator-like effector toolbox for genome engineering. Nature Protocols, 7: 171–192.

Segal, G.; Song, R.; Messing, J. 2003. A new opaque variant of maize by a single dominant RNA-interference-inducing transgene. Genetics, 165: 387–397.

Shimada, T.; Otani, M.; Hamada, T.; Kim, S.-H. 2006. Increase of amylose content of sweet potato starch by RNA interference of the starch branching enzyme II gene (*IbSBEII*). Plant Biotechnology, 23: 85–90.

Souza, A.J.; Mendes, B.M.J.; Filho, F.A.A.M. 2007. Gene silencing: concepts, applications, and perspectives in wood plants. Scientia Agricola, 64: 645–656.

Sunilkumar, G.; Campbell, L.M.; Puckhaber, L.; *et al.* 2006. Engineering cottonseed for use in human nutrition by tissue-specific reduction of toxic gossypol. Proceedings of the National Academy of Sciences U.S.A., 103: 18054–18059.

Tang, G.; Galili, G. 2004. Using RNAi to improve plant nutritional value: from mechanism to application. TRENDS in Biotechnology, 22 (9): 463–469.

Tang, G.; Galili, G.; Zhuang, X. 2007. RNAi and microRNA: breakthrough technologies for the improvement of plant nutritional value and metabolic engineering. Metabolomics 3: 357–369.

Tinoco, M.L.P.; Dias, B.B.A.; Dall'astta, R.C.; *et al.* 2010. In vivo trans-specific gene silencing in fungal cells by in planta expression of a double-stranded RNA. BMC Biology, 8: 27.

Todd, J.J.; Vodkin, L.O. 1993. Pigmented soybean (glycine-max) seed coats accumulate proanthocyanidins during development. Plant Physiology, 102: 663–670.

Todd, J.J.; Vodkin, L.O. 1996. Duplications that suppress and deletions that restore expression from a chalcone synthase multigene family. Plant Cell, 8: 687–699.

Tolia, N.H.; Joshua-Tor, L. 2006. Slicer and the Argonautes. Nature Chemical Biology, 3: 36–43.

Tomilov, A.A.; Tomilova, N.B.; Wroblewski, T.; *et al.* 2008. Trans-specific gene silencing between host and parasitic plants. Plant Journal, 56: 389–397.

Tu, H.M.; Godfrey, L.W.; Sun, S.S. 1998. Expression of the Brazil nut methionine-rich protein and mutants with increased methionine in transgenic potato. Plant Molecular Biology, 37: 829–838.

Tuteja, J.H.; Clough, S.J.; Chan, W.C.; Vodkin, L.O. 2004. Tissue-specific gene silencing mediated by a naturally occurring chalcone synthase gene cluster in *glycine max*. Plant Cell, 16: 819–835.

Tuteja, J.H.; Vodkin, L.O. 2008. Structural features of the endogenous CHS silencing and target loci in the soybean genome. Crop Science, 48: 49–69.

Tuteja, J.H.; Zabala, G.; Varala, K.; *et al.* 2009. Endogenous tissue-specific short interfering RNAs silence the chalcone synthase gene family in *glycine max* seed coats. The Plant Cell, 21: 3063–3077.

Upadhyay, S.K.; Chandrashekar, K.; Thakur, N.; *et al.* 2011. RNA interference for the control of whiteflies (*Bemisia tabaci*) by oral route. Journal of Biosciences, 36: 153–161.

Vanderschuren, H.; Stupak, E.; Fütterer, M.; *et al.* 2007. Engineering resistance to geminiviruses—review and perspectives. Plant Biotechnology Journal, 4: 1–14.

Vanderschuren, H.; Alder, A.; Zhang, P.; Gruissem, W. 2009. Dose-dependent RNAi-mediated geminivirus resistance in the tropical root crop cassava. Plant Molecular Biology, 70: 265–272.

Vanderschuren, H.; Moreno, I.; Anjanappa, R. B.; *et al.* 2012. Exploiting the combination of natural and genetically engineered resistance to cassava mosaic and cassava brown streak viruses impacting cassava production in Africa. PLoS One, 7: e45277.

Vazquez, F.; Vaucheret, H.; Rajagopalan, R.; *et al.* 2004. Endogenous trans-acting siRNAs regulate the accumulation of Arabidopsis mRNAs. Molecular Cell, 16: 69–79.

Vedova, C.B.D.; Lorbiecke, R.; Kirsch, H.; *et al.* 2005. The dominanti chalcone synthase allele *C2-Idf* (Inhibitor diffuse) from *Zea mays* (L.) acts via an endogenous RNA silencing mechanism, Genetics, 170: 1989–2002.

Wang, M.B.; Abbott, D.C.; Waterhouse, P.M. 2000. A single copy of a virus-derived transgene encoding hairpin RNA gives immunity to barley yellow dwarf virus. Molecular Plant Pathology, 6: 347–356.

Wang, Y.; Zhang, H.; Li, H.; Miao, X. 2011. Second-generation sequencing supply an effective way to screen RNAi targets in large scale for potential application in pest insect control. PLoS One, 6: e18644.

Waterhouse, P.M.; Graham, M.W.; Wang, M.B. 1998. Virus resistance and gene silencing in plants can be induced by simultaneous expression of sense and antisense RNA. Proceedings of the National Academy of Sciences U.S.A., 95: 13959–13964.

Warthmann, N.; Chen, H.; Ossowsk, S.; *et al.* 2008. Highly specific gene silencing by artificial miRNAs in rice. PLoS ONE 3: e1829.

Xei, Z.; Johansen, L. K.; Gustafson, A. M.; *et al.* 2004. Genetic and functional diversification of small RNA pathways in plants. Public Library of Science Biology, 2: 642–652.

Yadav, R.K.; Chattopadhyay, D. 2011. Enhanced viral intergenic region–specific short interfering RNA accumulation and DNA methylation correlates with resistance against a geminivírus. Molecular Plant-Microbe Interactions, 24: 1189–1197.

Yang, Y.; Sherwood, T.A.; Hiebert, C.P.; Polston, J.E. 2004. Use of *Tomato yellow leaf curl virus* (TYLCV) *rep* gene sequences to engineer TYLCV resistance in tomato. Phytopathology, 94: 490–496.

Yin, D.; Deng, S.; Zhan, K.; Cui, D. 2007. High-oleic peanut oils produced by hprna-mediated gene silencing of oleate desaturase. Plant Molecular Biology Reporter, 25: 154–163.

Zhu, X.; Galili, G. 2003. Increased lysine synthesis coupled with a knockout of its catabolism synergistically boostslysine contente and also transregulates the metabolism of other amino acids in Arabidopsis seeds. Plant Cell, 15: 845–853.

Zhu, X.; Galili, G. 2004. Lysine metabolism is concurrently regulated by synthesis and catabolism in both reproductive and vegetative tissues. Plant Physiology, 135: 129–136.

Index

Omics in Plant Breeding, First Edition. Edited by Aluízio Borém and Roberto Fritsche-Neto.
© 2014 John Wiley & Sons, Inc. Published 2014 by John Wiley & Sons, Inc.